Basics to production and

manufacturing of alcohol

Basics to the elaboration of ethanol, gasohol, E10, E20 and E85 fuels.

Theodore Ford, Alan Delfin

By reading this document, the reader agrees that under no circumstances is the author responsible for any losses, direct or indirect, which are incurred as a result of the use of information contained within this document, including, but not limited to, —errors, omissions, or inaccuracies.

Dedicatory

Dedicated to best friend, Víctor Manuel López Yépiz, who had supported me to accomplish my goals...

PS: What an asshole.

Table of Contents

INTRODUCTION

Alcohol (ethanol or ethyl alcohol) is the ingredient found in beer, wine and spirits that causes drunkenness.

Alcohol is formed when yeast ferments (breaks down without oxygen) the sugars in different food. For example, wine is made from the sugar in grapes, beer from the sugar in malted barley (a type of grain), cider from the sugar in apples, vodka from the sugar in potatoes, beets or other plants.

Alcohol is classed as a 'sedative hypnotic' drug, which means it acts to depress the central nervous system at high doses. At lower doses, alcohol can act as a stimulant, inducing feelings of euphoria and talkativeness, but drinking too much alcohol at one session can lead to drowsiness, respiratory depression (where breathing becomes slow, shallow or stops entirely), coma or even death.

As well as its acute and potentially lethal sedative effect at high doses, alcohol has effects on every organ in the body and these effects depend on the blood alcohol concentration (BAC) over time.

Our definition of alcohol fuel is a nearly 100 percent alcohol with a tad of water in it -- not a blend of alcohol with gasoline. So ... why an alcohol fuel? And why not a blend of gasoline and alcohol?

There are several reasons why we chose an alcohol fuel. The first and probably most important one is that alcohol can be made by anyone, with a minimum of equipment. The knowledge necessary to make it can be obtained just by reading this book. As long as folks can grow certain plants, they can make alcohol fuel to run all or part of their power equipment. Dependence upon someone else to supply that fuel is no longer a problem or a threat. Second, alcohol is a good fuel, superior to gasoline in many ways: It can give extra power to certain engines, it is almost non-polluting compared to gasoline, it is safe and easy to handle. Third, the cost of conversion from gasoline to alcohol is inexpensive: For many engines it is merely an adjustment of the carburetor jets.

Why not a gasohol fuel? The problem is water. Water and alcohol are totally miscible liquids. That is, they mix in all proportions. Pure alcohol and gasoline are also miscible liquids. But water and gasoline are not. This means that an alcohol-and-gasoline blend must be almost free of water. To make a 200-proof alcohol on the farm would require expensive equipment and additional production expenditures. At this time, that added expense would price a fuel blend beyond reason. But alcohol of 167 proof (16.5% water) is as good a fuel as 200-proof (100%) alcohol and better than gasohol.

Really, it comes down to basic survival. Right now, the fuel shortage doesn't seem all that serious.

WHAT IS ALCOHOL

Alcohol is a li□uid produced by fermentation. Further processing produces alcoholic drinks such as beer, wine, cider and spirits.

Alcohol, any of a class of organic compounds characterized by one or more hydroxyl (?OH) groups attached to a carbon atom of an alkyl group (hydrocarbon chain). Alcohols may be considered as organic derivatives of water (H_2O) in which one of the hydrogen atoms has been replaced by an alkyl group, typically represented by R in organic structures. For example, in ethanol (or ethyl alcohol) the alkyl group is the ethyl group, $?CH_2CH_3$.

Methane, in which four hydrogen atoms are bound to a single carbon atom, is an example of a basic chemical compound. The structures of chemical compounds are influenced by complex factors, such as bond angles and bond length.

Chemical Composition Explanation

An oxygen atom normally forms two s bonds with other atoms; the water molecule, H_2O, is the simplest and most common example. If one hydrogen atom is removed from a water molecule, a hydroxyl functional group (OH) is generated.

Alcohols are among the most common organic compounds. They are used as sweeteners and in making perfumes, are valuable intermediates in the synthesis of other compounds, and are among the most abundantly produced organic chemicals in industry. Perhaps the two best-known alcohols are ethanol and methanol (or methyl alcohol). Ethanol is used in toiletries, pharmaceuticals, and fuels, and it is used to sterilize hospital instruments. It is, moreover, the alcohol in alcoholic beverages. The anesthetic ether is also made from ethanol. Methanol is used as a solvent, as a raw material for the manufacture of formaldehyde and special resins, in special fuels, in antifreeze, and for cleaning metals.

Alcohols may be classified as primary, secondary, or tertiary, according to which carbon of the alkyl group is bonded to the hydroxyl group. Most alcohols are colourless liquids or solids at room temperature. Alcohols of low molecular weight are highly soluble in water; with increasing molecular weight, they become less soluble in water, and their boiling points, vapour pressures, densities, and viscosities increase.

Ethanol: Formula and Structure

Did you know that the red fluid that rises in a thermometer is ethanol? Did you also know that the alcohol present in alcoholic beverages is ethanol?

Ethanol, also commonly referred to as ethyl alcohol, pure alcohol, grain alcohol, and drinking alcohol, is most known as the alcohol present in alcoholic beverages. Ethanol, which can also be abbreviated as EtOH, is a colorless liquid with a slight odor, and it is soluble in water. It is flammable and volatile, so it evaporates easily when left in an open container.

Ethanol's chemical formula is C_2H_6O. This chemical formula can also be written as CH_3CH_2OH or

C2H5OH. It is made of nine atoms that include two carbon (C) atoms, six hydrogen (H) atoms, and one oxygen (O) atom.

Its chemical structure is illustrated in the following picture. Here, there is a methyl group (which is the CH3-), a methylene group (which is the -CH2-), and a hydroxyl group (which is the -OH) in the chemical structure.

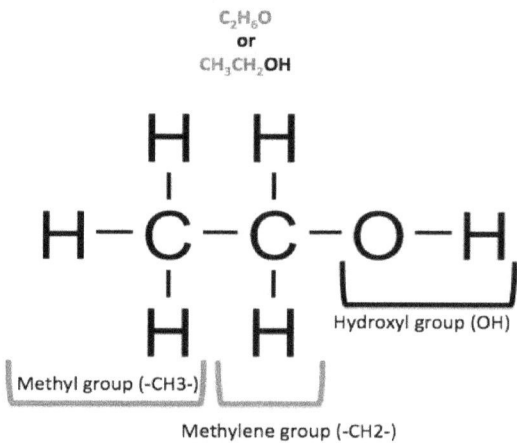

Ethanol Uses

Ethanol, also commonly referred to as ethyl alcohol, pure alcohol, grain alcohol, and drinking alcohol, is most known as the alcohol present in alcoholic beverages. Ethanol, which can also be abbreviated as EtOH, is a colorless liquid with a slight odor, and it is soluble in water. It is flammable and volatile, so it evaporates easily when left in an open container.

Ethanol's chemical formula is C_2H_6O. This chemical formula can also be written as CH_3CH_2OH or C_2H_5OH. It is made of nine atoms that include two carbon (C) atoms, six hydrogen (H) atoms, and one oxygen (O) atom.

Its chemical structure is illustrated in the following picture. Here, there is a methyl group (which is the CH_3-), a methylene group (which is the $-CH_2-$), and a hydroxyl group (which is the -OH) in the chemical structure.

As A Fuel (Why Is A Fuel)

Almost every country can become energy independent. Anywhere that has sunlight and land can produce alcohol from plants. Brazil, the fifth largest country in the world imports no oil, since half its cars run on alcohol fuel made from sugarcane, grown on 1% of its land.

We can reverse global warming. Since alcohol is made from plants, its production takes carbon dioxide out of the air, se□uestering it, with the result that it reverses the greenhouse effect (while potentially vastly improving the soil). Recent studies show that in a permaculturally designed mixed-crop alcohol fuel production system, the amount of greenhouse gases removed from the atmosphere by plants and then exuded by plant roots into the soil as sugar can be 13 times what is emitted by processing the crops and burning the alcohol in our cars.

We can revitalize the economy instead of suffering through Peak Oil. Oil is running out, and what we replace it with will make a big difference in our environment and economy. Alcohol fuel production and use is clean and environmentally sustainable, and will revitalize families, farms, towns, cities, industries, as well as the environment. A national switch to alcohol fuel would provide many millions of new permanent jobs.

No new technological breakthroughs are needed. We can make alcohol fuel out of what we have, where we are. Alcohol fuel can efficiently be made out of many things, from waste products like stale donuts, grass clippings, food processing waste-even ocean kelp. Many crops produce many times more alcohol per acre than corn, using arid, marshy, or even marginal land in addition to farmland. Just our lawn clippings could replace a third of the autofuel we get from the Mideast.

Unlike hydrogen fuel cells, we can easily use alcohol fuel in the vehicles we already own. Unmodified cars can run on 50% alcohol, and converting to 100% alcohol or flexible fueling (both alcohol and gas) costs only a few hundred dollars. Most auto companies already sell new dual-fuel vehicles.

Alcohol is a superior fuel to gasoline! It's 105 octane, burns much cooler with less vibration, is less flammable in case of accident, is 98% pollution-free, has lower evaporative emissions, and deposits no carbon in the engine or oil, resulting in a tripling of engine life. Specialized alcohol engines can get at least 22% better mileage than gasoline or diesel.

It's not just for gasoline cars. We can also easily use alcohol fuel to power diesel engines, trains, aircraft, small utility engines, generators to make electricity, heaters for our homes—and it can even be used to cook our food.

Alcohol has a proud history. Gasoline is a refinery's toxic waste; alcohol fuel is li uid sunshine. Henry Ford's early cars were all flex-fuel. It wasn't until gasoline magnate John D. Rockefeller funded Prohibition that alcohol fuel companies were driven out of business.

The byproducts of alcohol production are clean, instead of being oil refinery waste, and are worth more than the alcohol itself. In fact, they can make petrochemical fertilizers and herbicides obsolete. The alcohol production process concentrates and makes more digestible all protein and non-starch nutrients in the crop. It's so nutritious that when used as animal feed, it produces more meat or milk than the corn it comes from. That's right, fermentation of corn increases the food supply and lowers the cost of food.

Locally produced ethanol supercharges regional economies. Instead of fuel expenditures draining capital away to foreign bank accounts, each gallon of alcohol produces local income that gets recirculated many times. Every dollar of tax credit for alcohol generates up to $6 in new tax revenues from the increased local business.

Alcohol production brings many new small-scale business opportunities. There is huge potential for profitable local, integrated, small-scale businesses that produce alcohol and related byproducts, whereas when gas was cheap, alcohol plants had to be huge to make a profit.

Scale matters most of the widely publicized potential problems with ethanol are a function of scale. Once production plants get beyond a certain size and are too far away from the crops that supply them, closing the ecological loop becomes problematic. Smaller-scale operations can more efficiently use a wide variety of crops than huge specialized one-crop plants, and diversification of crops would largely eliminate the problems of monoculture.

The byproducts of small-scale alcohol plants can be used in profitable, energy-efficient, and environmentally positive ways.

 For instance, spent mash (the liquid left over after distillation) contains all the nutrients the next fuel crop needs and can return it back to the soil if the fields are close to the operation. Big-scale plants, because they bring in crops from up to 45 miles away, can't do this, so they have to evaporate all the water and sell the resulting byproduct as low-price animal feed,which accounts for half the energy used in the plant.

Advantages of using alcohol as a fuel

It can be obtained from both natural and manufacturing methods.

It has a very high octane number (above 100) and a very good flame speed.

It produces less emissions than gasoline (petrol).

Due to cooler intake, volumetric efficiency of engine running on alcohol is good.

Alcohol has low sulphur content in fuel.

More moles of exhaust gases produced provide more power to expansion stroke.

GASOHOL AND E85

Gasohol is an alternative fuel consisting of a mixture of typically 90 percent gasoline with 10 percent anhydrous ethanol. Gasohol can be used in most modern and light duty vehicles with an internal combustion. The Gasohol blend of 90 percent gasoline and 10 percent anhydrous ethanol has been approved for use in several countries and can be used with no modification to the vehicle's engine.

Gasohol is useful in decreasing the population's dependence on foreign oil, and reduces the carbon monoxide emissions by up to 30 percent.

The ethanol typically used in the Gasohol production is derived from fermenting agricultural crops. Fuel containing ethanol normally has an "E" number which explains the mixture. E10 consists of 10 percent ethanol and 90 percent gasoline whereas E85 is a blend of 85 percent ethanol and 15 percent gasoline. E5 and E7 are also common ethanol blends.

Ethanol blends are used in the following countries:

E10 - Australia

E10 - Austria

E20 to E25 - Brazil

E5/E10 - Canada

E10 - China

E10 - Colombia

E7 - Costa Rica

E5 - Denmark

E5 - Finland

E5 - India

E10 - Jamaica

E10 - New Zealand

E10 - Pakistan

E12 - Paraguay

E5 - Sweden

E10/E20 - Thailand

E10 - United States (in 10 states)

Blends of Gasohol

E5 to E25 are known as low ethanol blends, and have 5 to 25 percent ethanol blended with 95 to 75 percent gasoline.

E30 to E85 are considered to be high ethanol blends and have 30 to 85 percent ethanol mixed with 70 to 15 percent gasoline.

The most popular Gasohol blend is E10, which consists of 10 percent ethanol and 90 percent gasoline due to the fact that no modifications are needed to a vehicle's engine to use E10.

Gasohol Advantages

In a search for alternative fuel options the usage of Gasohol has certain advantages:

Emissions from using Gasohol are less than that of vehicles using gasoline. Emissions are not only harmful to the environment but can cause serious problems and even death in humans. By minimizing the emissions expelled into the atmosphere, we not only ensure a greener environment, but also a physically healthier population eliminating illness. Such as, asthma and heart disease, caused by vehicle emissions.

Using Gasohol assists in the reduction of oil imported from other countries. Not only does this lessen our carbon footprint but, with Gasohol production of up to 85 percent ethanol, less oil needs to be imported to manufacture gasoline.

Crop prices are raised with the production of Gasohol. Ethanol is an alcohol derived from crops such as cane, grains and sorghum. This increases the demand and ultimately the price of these crops.

Gasohol is typically cheaper than petroleum as it is cheaper to manufacture.

With most of the world's automobiles running on Gasohol, it is only a matter of time before vehicles capable of running on pure ethanol will be designed.

Although not the perfect solution, Gasohol is a step in the right direction to finding alternative fuels for our automobiles. The positive aspects of Gasohol outweigh the negatives and, with more countries making Gasohol available at their gas stations, everybody is making an effort in ensuring less dangerous gasses are expelled into the atmosphere.

Types Of E10, E15, E20 ,E85

(Brazil Had Been Using Many Mixtures Thru The Years , Explain Further.)

An E85 engine is a gasoline engine with some parts changed to be non-corrosive to ethanol. Since ethanol contains less energy per gallon, burning it in a low compression ratio gasoline engine results in very poor mileage 75% of that of gasoline.

If you go 300 miles on a tank of gasoline, and fill it with E85, you will only go 225 miles. For every 3 stops to fill-up with gasoline, the driver will have to stop 4 times if running E85.

The current process is that the ethanol producer ships the ethanol to the oil company blender who adds gasoline to it to make either E10 or E85.

This process means that the oil companies control the price of E85 just like they control the price of gasoline.

E85 is priced so that it costs the consumer more money to burn E85 per mile than gasoline, roughly $400 per year according to www.fueleconomy.gov. The ethanol industry does control the price of its product to the final consumer.

E85 began life due to a government mandate, the Corporate Average Fuel Economy Act (CAFE). It has been in the marketplace for 25 years, since 1992, with 100's of millions of dollars of advertising put into it. It has gained only 0.2% market share penetration (306 million gals over 143 billion gals of gasoline) after all that time and money.

Consumers have plainly indicated they aren't buying this product. The vast majority of gasoline retailers are franchisees, not owned by the oil companies. They are very reluctant to purchase E85 and put it into one of their tanks since they believe it won't sell. They are right.

This is why, as of December 2018, there are only 4,498 E85 stations in the US (E85 Prices, RFA), not because of any distribution problems or lack of E85 capable vehicles.

Another way to look at this is: If E85 offered the same mileage as gasoline, we would have switched over to it long ago since it costs less per gallon, and motor fuel is a very price sensitive product.

Congressional bills have been introduced to re□uire a rapid increase in the number of new E85 capable vehicles. This does not seem a reasonable strategy to force the manufacture of more E85 vehicles when the public has clearly shown over the last 20 years that it isn't interested in buying E85.

E15, E20, E30

There is a lot of consideration being given to intermediate blends such as E15, E20, and E30 and the EPA has approved the sale of E15 for certain model year vehicles and a few stations in the U.S. are selling it.

There is a good case for E15 in that it would immediately help the ethanol industry which is having trouble growing due to the informal blend limit of 10% ethanol.

However, there is a real dispute about potential problems with storage tanks, piping, and pumps. Also, there is no requirement that E15 actually be used and even if it comes into use, it cannot make us independent of imported oil or significantly lower CO_2 emissions.

Food versus fuel is the dilemma regarding the risk of diverting farmland or crops for biofuels production to the detriment of the food supply. The biofuel and food price debate involves wide-ranging views, and is a long-standing, controversial one in the literature. There is disagreement about the significance of the issue, what is causing it, and what can or should be done to remedy the situation. This complexity and uncertainty is due to the large number of impacts and feedback loops that can positively or negatively affect the price system. Moreover, the relative strengths of these positive and negative impacts vary in the short and long terms, and involve delayed effects. The academic side of the debate is also blurred by the use of different economic models and competing forms of statistical analysis.

Biofuel production has increased in recent years. Some commodities like maize (corn), sugar cane or vegetable oil can be used either as food, feed, or to make biofuels. For example, since 2006, a portion of land that was also formerly used to grow other crops in the United States is now used to grow corn for biofuels, and a larger share of corn is destined to ethanol production, reaching 25% in 2007. Second generation biofuels could potentially combine farming for food and fuel and moreover, electricity could be generated simultaneously, which could be beneficial for developing countries and rural areas in developed countries. With global demand for biofuels on the increase due to the oil price increases taking place since 2003 and the desire to reduce oil dependency as well as reduce GHG emissions from transportation, there is also fear of the potential destruction of habitats by being converted into farmland. Environmental groups have raised concerns about this trade-off for several years, but the debate reached a global scale due to the 2007–2008 world food price crisis. On the other hand, several studies do show that biofuel production can be significantly increased without increased acreage. Therefore, stating that the crisis in hand relies on the food scarcity.

Biofuels are not a new phenomenon. Before the industrialisation, horses were the primary (and humans probably the secondary) source of power for transportation and physical work, re□uiring food. The growing of crops for horses (typically oat) for carrying out physical work is of course comparable to the growing of crops for biofuels for engines, albeit on a smaller scale, because production since then has increased.

Ethanol fuel as an oxygenate additive

The demand for ethanol fuel produced from field corn was spurred in the U.S. by the discovery that methyl tertiary butyl ether (MTBE) was contaminating groundwater. MTBE use as an oxygenate additive was widespread due to mandates of the Clean Air Act amendments of 1992 to reduce carbon monoxide emissions. As a result, by 2006 MTBE use in gasoline was banned in almost 20 states. There was also concern that widespread and costly litigation might be taken against the U.S. gasoline suppliers, and a 2005 decision refusing legal protection for MTBE, opened a new market for ethanol fuel, the primary substitute for MTBE. At a time when corn prices were around US$2 a bushel, corn growers recognized the potential of this new market and delivered accordingly. This demand shift took place at a time when oil prices were already significantly rising.

Oil price increases

Oil price increases since 2003 resulted in increased demand for biofuels. Transforming vegetable oil into biodiesel is not very hard or costly so there is a profitable arbitrage situation if vegetable oil is much cheaper than diesel. Diesel is also made from crude oil, so vegetable oil prices are partially linked to crude oil prices. Farmers can switch to growing vegetable oil crops if those are more profitable than food crops. So all food prices are linked to vegetable oil prices, and in turn to crude oil prices. A World Bank study concluded that oil prices and a weak dollar explain 25–30% of total price rise between January 2002 until June 2008.

Demand for oil is outstripping the supply of oil and oil depletion is expected to cause crude oil prices to go up over the next 50 years. Record oil prices are inflating food prices worldwide, including those crops that have no relation to biofuels, such as rice and fish.

In Germany and Canada it is now much cheaper to heat a house by burning grain than by using fuel derived from crude oil. With oil at $120/barrel a savings of a factor of 3 on heating costs is possible.

When crude oil was at $25/barrel there was no economic incentive to switch to a grain fed heater.

Livestock Production Vs Energy Crops

Globally, there is an increasing demand for land for livestock production (grazing and animal feed production), driven partly by rising middle class incomes in rapidly expanding economies. Not all land used for meat production is essential for human nutrition, with much consumption being a lifestyle choice. Production of plant protein is less energy intensive. Some land currently used for livestock could potentially be used for energy crops, without a negative impact on human health or any increase in GHG emissions. Some reports suggest that conversion of land from meat production to bioenergy feedstocks could offer environmental and health benefits, for example Zero Carbon Britain.

In September 2013, the FAO report Tackling Climate Change through Livestock estimated that 7.1 gigatonnes of carbon dioxide e□uivalent (CO2-e□) per annum was produced by the livestock industry representing 14.5 % of all human-induced emissions (other studies put the figue as high as 18 %, more than the total emissions from all forms of transport). Beef and dairy production account for the majority of emissions, respectively contributing 41 and 19 % of the sector's emissions; while pig meat and poultry meat and eggs contribute respectively 9% and 8% to the sector's emissions. Sources of emissions include : feed production and processing (45% of the total – with 9 percent attributable to the expansion of pasture and feed crops into forests), enteric fermentation from ruminants (39%), and manure decomposition (10%). The remainder is attributable to the processing and transportation of animal products.

LEGAL

Getting a permit (US)

The Ultimate Guide

In order to legally make fuel alcohol one must have the proper permits. A federal fuel alcohol producer permit is available for free from the TTB. The application must be submitted and approved by the TTB before fuel alcohol is manufactured. The TTB is the division of the government that is in charge of regulating and overseeing distillation. TTB stands for Alcohol and Tobacco Tax and Trade Bureau which is part of the U.S. Department of the Treasury.

Even though the Federal Government allows the manufacture of fuel alcohol with a permit it is important to realize that some states don't allow fuel alcohol production. State distilling laws vary from state to state. Some states have no laws on owning a still, but prohibit the distillation of alcohol while other states prohibit possession of a still unless it's for fuel alcohol. States such as as North Carolina re uires a state fuel alcohol permit as well as the federal fuel permit. Some states may prohibit possession of distillation e□uipment and distilling altogether. It is important to research local state laws before applying for a federal fuel permit.

Download the federal fuel permit here and follow along with our guide to properly fill out the alcohol fuel plant application. Our guide will focus on the small fuel alcohol production plant as production with a Clawhammer Supply, LLC still will produce less than 10,000 proof gallons a year.

1)Type Of Plant: If using a Clawhammer Supply still check the Small – 10,000 Proof Gallons or Less box.

2) Amended Permit: Leave this blank when applying for a new permit

3)Name of Owner: Type in first / middle/ and last name

4) Daytime Telephone Number: Enter full phone number

5) EIN or SSN- Enter your Social Security Number or if registered as a business use the EIN- Employer Identification Number

6) DOB: Enter the DOB for the applicants or applicants.

7) Location: Type in the street name where the small distillation plant will be located.

8) Mailing address: If the mailing address is different than the plant location enter it here. If the mailing address is the same as the plant location leave it blank.

9) Premise for Alcohol Fuel Plant Are: If you own the premise check owned by the applicant and skip to number 11.

10) If the premise is rented fill out section 10 which must be signed by the property owner.

11) Stills For Fuel Production On Plant Premises:

a) Still Manufacturer: Clawhammer Supply, LLC

b) Serial Number: The Serial Number for Clawhammer Supply, LLC is the order number

c) Kind of Still: Pot Still

d) Capacity – The Clawhammer Supply, LLC still is a pot still and a pot still may be shown by giving the wine gallon capacity of the pot still.

a. The 1 gallon Clawhammer Supply, LLC still 1 gallon

b. The 5 gallon Clawhammer Supply, LLC still 5 gallons

c. The 10 gallon Clawhammer Supply, LLC still 10 gallons

12) Basic Materials to be used in production of spirits: If you are planning on using a variety of fermentables select the appropriate boxes. I selected grain, sugar based crops, and fruit or fruit products as these are the fermentables I use when making a mash.

13) Security Measures: On our property we have a fenced in yard and a distilled spirits building which is locked.

14) Diagram of plant premise: The diagram of your plant premises may be drawn by hand and does not have to be drawn to scale.

15) I will comply with the clean water act: Check I will comply with the clean water act and don't discharge

into navigable water s of the U.S. Navigable waters are waterways that provide a channel for commerce and transportation of people and goods.

16) Skip: Only applies to Medium and Large Alcohol fuel plant applications

17) Sign the form

18) Title: Owner

19) Enter the date

Print out 3 copies of the TTB application and any attachments (the drawing of the premise); make sure the attachments are labeled with first and last name and are printed on the same size paper as the application. Mail the completed application to: Director, National Revenue Center, 550 Main St Ste 8002, Cincinnati, OH 45202-5215. If re□uired by your state, submit a copy of your approved application to the alcohol beverage agency or other State agency. It took 3 months to receive fuel alcohol permit.

Safety

The greatest danger from the use of gasoline as a vehicle fuel is from fires. Gasoline fires in vehicles result in hundreds of deaths and millions of dollars in property damage each year. Methanol does not evaporate or form vapors as readily as gasoline does, and methanol vapors must be four times more concentrated in air than gasoline to ignite. Methanol burns 75% slower than gasoline, and methanol fires release heat at only one-eighth the rate of gasoline. Unlike gasoline fires, methanol fires can be extinguished with water. Methanol is inherently more difficult to ignite than gasoline, and much less likely to cause deadly or damaging car fires if it does ignite. The U.S. EPA has estimated that if all our cars were fueled with methanol, the incidence of vehicle fires would be reduced by 90%, saving hundreds of lives per year.

Methanol, like gasoline or diesel fuels, should never be ingested and is toxic. Deaths have been reported from intake of as little as 13 ml. of gasoline (less than one ounce), which is similar to the fatal ingestion range for methanol. Our bodies contain methanol naturally, and it is found in many parts of our diet, including fresh fruit, vegetables, and fermented foods and beverages. Both methanol and gasoline can be absorbed through human skin, and the response for both is the same: remove any contaminated clothing, and wash with soap and water. Spill-free methanol nozzles have been developed that will prevent a consumer from even being able to come into contact with methanol fuel. During California's decade long methanol fuel program, there was not a single case of methanol poisoning from the operation of thousands of methanol vehicles traveling a cumulative 200 million miles. Similarly, over 200 million miles of methanol vehicle operation has been accumulated in China without a single reported case of methanol poisoning. Quite simply, we don't drink our fuels.

Methanol Safety Tips

Methanol is a natural by-product of wood alcohol, natural gas and coal. It has also recently garnered attention for being used in biofuel. But don't let the "green" of this chemical fool you. Methanol is very dangerous and can cause death when mishandled. Therefore, it is crucial to understand how to handle methanol in a safe and effective way.

Hazards Associated with Using Methanol

Methanol is highly flammable and toxic. Direct ingestion of more than 10mL can cause permanent blindness by destruction of the optic nerve, poisoning of the central nervous system, coma and possibly death. These hazards are also true if methanol vapors are inhaled.

When handling methanol, including biofuels, it is best to avoid direct exposure as much as possible. As such, it is imperative that safety gear be worn, especially those that cover the face, eyes and skin. If working where methanol vapors are present, proper ventilation is imperative for safety.

Should methanol come into direct contact with the skin, remove any contaminated clothing and wash the affected area with soap and water for 15 minutes. If methanol comes into contact with the eyes, flush immediately with tepid water for 15 minutes and then seek qualified medical help. Lastly, if ingested, seek immediate medical attention and do not induce vomiting.

If a methanol spill occurs, stop or reduce the discharge of the methanol (if it can be done without risk) and immediately call your local fire department. Try to isolate the spill/leak area in at least 330 to 660 feet in all directions. Remove and eliminate all sources of potential ignition and stay upwind. Do not touch or walk through any spillage and prevent the spilled methanol from entering into waterways, sewers, basements or confined areas.

When storing methanol, it should always be kept within a closed system or environmentally-approved container; never leave methanol open to the air. Label containers in accordance with local regulations and site requirements. Comprehensive product handling procedures and systems need to be in place at all storage and transfer points. Methanol is mostly non-corrosive when stored with metals at ambient temperatures, but are corrosive with lead, magnesium or platinum.

Safety Tips When Disposing of Methanol

When seeking to dispose of methanol, large ⬜uantities of waste can be disposed of at a licensed waste solvent company or reclaimed by filtration and distillation. Waste methanol, or water contaminated with methanol, must never be put directly into sewers or surface waters, nor poured down the drain, on the ground or into any body of water.

In the United States, the fire code (also fire prevention code or fire safety code) is a model code adopted by the state or local jurisdiction and enforced by fire prevention officers within municipal fire departments. It is a set of rules prescribing minimum re□uirements to prevent fire and explosion hazards arising from storage, handling, or use of dangerous materials, or from other specific hazardous conditions. It complements the building code. The fire code is aimed primarily at preventing fires, ensuring that necessary training and e□uipment will be on hand, and that the original design basis of the building, including the basic plan set out by the architect, is not compromised. The fire code also addresses inspection and maintenance re□uirements of various fire protection e□uipment in order to maintain optimal active fire protection and passive fire protection measures.

A typical fire safety code includes administrative sections about the rule-making and enforcement process, and substantive sections dealing with fire suppression e□uipment, particular hazards such as containers and transportation for combustible materials, and specific rules for hazardous occupancies, industrial processes, and exhibitions.

Sections may establish the re□uirements for obtaining permits and specific precautions re□uired to remain in compliance with a permit. For example, a fireworks exhibition may re□uire an application to be filed by a licensed pyrotechnician, providing the information necessary for the issuing authority to determine whether safety re□uirements can be met. Once a permit is issued, the same authority (or another delegated authority) may inspect the site and monitor safety during the exhibition, with the power to halt operations, when unapproved practices are seen or when unforeseen hazards arise.

Preventing and Controlling

Ethanol Fires

• Alternative fuel, new focus on danger

• Gasoline fires vs. Ethanol fires

• Solubility in water / Specific gravity

- Conductivity / Vapor density / Toxicity

- Fires, public safety

- Spills / Small fires / Big fires

New Focus on Danger

- E85 is highly flammable, and will be easily ignited by heat, sparks or flames.

- E85 is a polar/water-miscible flammable (i.e., they mix readily with water)

- Flame visibility: A fuel ethanol flame is less bright than a gasoline flame but is easily visible in daylight.

Gasoline fires vs. Ethanol fires

- Foam is used to blanket the top of burning gasoline and usually snuffs out of the flames.

- Ethanol fires re□uire a special alcohol - resistant foam that relies on long-chain molecules known as polymers to smother the flames.

Solubility in water / Specific gravity

• Solubility in water: Fuel ethanol will mix with water, but at high enough concentrations of water, the ethanol will separate from the gasoline.

• Specific gravity: Pure ethanol and ethanol blends are heavier than gasoline.

Conductivity

• Conductivity: Ethanol and ethanol blends conduct electricity. Gasoline, by contrast, is an electrical insulator.

Vapor Density / Toxicity

• Vapor density: Ethanol vapor, like gasoline vapor, is denser than air and tends to settle in low areas. However, ethanol vapor disperses rapidly.

• Toxicity: Ethanol is less toxic than gasoline or methanol. Carcinogenic compounds are not present in pure ethanol; however, because gasoline is used in the blend, E85 is considered to be potentially carcinogenic.

Good News about Ethanol

• Flammability: At low temperature (32°), E85 vapor is more flammable thangasoline vapor. However at normal temperatures, E85 vapor is less flammable than gasoline.

Spills

• ELIMINATE all ignition sources (no smoking, flares, sparks or flames in immediate area).

• Do not touch or walk through spille material.

• Stop leak if you can do it without risk.

• Use clean non-sparking tools to collect absorbed material.

Fires, public safety

• Call 911 immediately

- Keep unauthorized personnel away

- Stay upwind

- Keep out of low areas

- Structural firefighters' protective clothing will only provide limited protection

Extinguishing Small Ethanol Fires

- Use a CO_2, halon, or dry chemical extinguisher that is marked B, C, BC, or ABC.

- An alcohol An alcohol type or alcohol type or alcohol-resistant (ARF) resistant (ARF) foam may be used to effectively combat fuel ethanol fires.

- Never use water to control a fire involving high-concentration fuel ethanol such as E85

Firefighters will:

- Fight fire from maximum distance

- Cool containers with flooding quantities of water until well after fire is out

• Withdraw immediately in case of rising sound from venting safety devices or discoloration of tank

• ALWAYS stay away from tanks engulfed in fire

The oil and gas industry is a dynamic marketplace, full of economic highs and lows and more than its share of physical dangers. As with any specialized industry, there are uni☐ue risks in the oil and gas business, and risk management is at the core of protecting business assets. U.S. Risk Underwriters, a prominent special-risks insurance company, works with insurance agents and brokers across the country to provide comprehensive insurance solutions for oil and gas industry clients. The firm has extensive experience in navigating the challenges of the industry, and offers insurance programs that meet the needs of industry clients perfectly.

Risks in the Oil & Gas Industry

As of 2015, the United States has become the leading producer of oil and gas in the world. Even as prices plummeted and drilling rigs were discontinued in favor of advanced technologies like hydraulic fracturing, production increased dramatically. Drilling rig count, or the number of active drilled wells, was a common risk metric for insurers. Now that other technologies have taken traditional drilling's place, this can no longer be relied upon as an accurate measure of production capability.

Along with new technologies come new risks, particularly in the case of hydraulic fracturing, or "fracking". Wells drilled in one insurance policy period may continue to produce years after the initial well placement, creating significant issues for insurance brokers and their clients. Physical risks, such as workplace accidents, environmental contamination, and blowouts compound the inherent dangers in this industry. In 2017, accident rates experienced a downturn, thanks to improved safety standards and automation of critical oil/gas production tasks. However, when an accident does occur, the insurance claims tend to be substantially higher than in years past. This is partially due to the complicated and expensive technologies employed to maximize production levels, and higher claim levels can also be attributed to emerging energy markets. A single oilfield accident, such as a catastrophic blowout or explosion, can cost millions of dollars in cleanup expenses, lost business, and worker injury claims.

Negligence or professional liability claims may also cost thousands or even millions of dollars in unforeseen expenses for production companies.

Faced with these risks presented above, it is clear that companies in the oil and gas production industry need insurance solutions. No matter the operation, mitigating risks is part of the overall oil and gas industry picture, and insurance plays an important role in risk management. Clients of insurance agents may be oilfield operators and well-servicing contractors, consultants, equipment sales and rental operations, and manufacturers, to name only a few of the many players in this dynamic energy production industry. Each of those players must balance potential risks against the insurance products available to protect their financial interests.

Specialized Insurance Solutions for the Oil & Gas Industry

U.S. Risk Underwriters has many years of experience in serving oil and gas production clients of partner insurance agents. The firm has developed its own Energy & Environmental division to better address the insurance risks production companies face. Insurance solutions provided by the company are known for their comprehensive coverage and flexibility, ensuring that the specific needs and risks of each client are covered. A range of coverage lines are available, including general (GL) and professional liability (PL), inland marine, contractors pollution liability (CPL), and environmental impact/impairment liability (EIL) or Site Pollution. This last component is critical, as typical liability policies specifically exclude pollution as a covered risk, a problem that has plagued the industry for many years. Each of these solutions is tailored to clients in both upstream and downstream operations. With the right insurance products in place, the expenses of oilfield worker injuries or a catastrophic e uipment failure can be managed successfully, protecting the valuable business assets of the insured.

Batches record keeping

Documentation is the key to GMP compliance and ensures traceability of all development, manufacturing, and testing activities. Documentation provides the route for auditors to assess the overall quality of operations within a company and the final product.

The management of each operational site is required to define responsibility for origination, distribution, maintenance, change control, and archiving of all GMP documentation and records within that department or unit.

Document owners are required to ensure that all aspects of documentation and records management specified in form of standard operating procedures (SOPs).

All associates have the responsibility of ensuring that all GMP activities are performed according to the official SOPs; any deviations in procedure are reported to their supervisor and are adequately documented.

The local Quality assurance unit has the responsibility of ensuring via organizational measures and auditing that GMP documentation and records systems used within the operational unit are complete and comply with the relevant GMP reQuirements, and also that the requirements of the SOPs are followed.

ReQuirements for specific documents or record, including ownership, content, authorization, and change control procedures, has to be described or cross-referenced in the Quality modules which relate to the subject of the document.

General reQuirements

Good documentation constitutes an essential part of the Quality assurance system. Clearly written procedures prevent errors resulting from spoken communication, and clear documentation permits tracing of activities performed.

Documents must be designed, prepared, reviewed, and distributed with care.

Documents must be approved, signed, and dated by the appropriate competent and authorized persons.

Documents must have unambiguous contents. The title, nature, and purpose should be clearly stated. They must be laid out in an orderly fashion and be easy to check. Reproduced documents must be clear and legible.

Documents must be regularly reviewed and kept up-to-date. When a document has been revised, systems must be operated to prevent inadvertent use of superseded documents (e.g., only current documentation should be available for use).

Documents must not be handwritten; however, where documents re□uire the entry of data, these entries may be made in clear legible handwriting using a suitable indelible medium (i.e., not a pencil). Sufficient space must be provided for such entries.

Any correction made to a document or record must be signed or initialed and dated; the correction must permit the reading of the original information. Where appropriate, the reason for the correction must be recorded.

Record must be kept at the time each action is taken and in such a way that all activities concerning the conduct of preclinical studies, clinical trials, and the manufacture and control of products are traceable.

Storage of critical records must at secure place, with access limited to authorized persons. The storage location must ensure ade□uate protection from loss, destruction, or falsification, and from damage due to fire, water, etc.

Records which are critical to regulatory compliance or to support essential business activities must be duplicated on paper, microfilm, or electronically, and stored in a separate, secure location in a separate building from the originals.

Date may be recorded by electromagnetic or photographic means, but detailed procedures relating to whatever system is adopted must be available. Accuracy of the record should be checked as per the defined procedure. If documentation is handled by electronic data processing methods, only authorized persons should be able to enter or modify data in the computer, access must be restricted by passwords or other means, and entry of critical data must be independently checked.

It is particularly important that during the period of retention, the data can be rendered legible within an appropriate period of time.

If data is modified, it must be traceable.

There are various types of procedures that a GMP facility can follow. Given below is a list of the most common types of documents, along with a brief description of each.

Quality manual: A global company document that describes, in paragraph form, the regulations and/or parts of the regulations that the company is re☐uired to follow.

Policies: Documents that describe in general terms, and not with step-by-step instructions, how specific GMP aspects (such as security, documentation, health, and responsibilities) will be implemented.

Standard operating procedures (SOPs): Step-by-step instructions for performing operational tasks or activities.

Batch records: These documents are typically used and completed by the manufacturing department. Batch records provide step-by-step instructions for production-related tasks and activities, besides including areas on the batch record itself for documenting such tasks.

Test methods: These documents are typically used and completed by the quality control (QC) department. Test methods provide step-by-step instructions for testing supplies, materials, products, and other production-related tasks and activities, e.g., environmental monitoring of the GMP facility.

Test methods typically contain forms that have to be filled in at the end of the procedure; this is for documenting the testing and the results of the testing.

Specifications: Documents that list the re□uirements that a supply, material, or product must meet before being released for use or sale. The QC department will compare their test results to specifications to determine if they pass the test.

Logbooks: Bound collection of forms used to document activities. Typically, logbooks are used for documenting the operation, maintenance, and calibration of a piece of e□uipment. Logbooks are also used to record critical activities, e.g., monitoring of clean rooms, solution preparation, recording of deviation, change controls and its corrective action assignment.

Denaturalizing alcohol to be used as fuel

Denatured alcohol, also called methylated spirit (in Australia, New Zealand, South Africa and the United Kingdom) or denatured rectified spirit, is ethanol that has additives to make it poisonous, bad-tasting, foul-smelling, or nauseating to discourage recreational consumption. It is sometimes dyed. Pyridine, methanol, or both can be added to make denatured alcohol poisonous, and denatonium can be added to make it bitter.

Denatured alcohol is used as a solvent and as fuel for alcohol burners and camping stoves. Because of the diversity of industrial uses for denatured alcohol, hundreds of additives and denaturing methods have been used. The main additive has traditionally been 10% methanol, giving rise to the term "methylated spirits". Other typical additives include isopropyl alcohol, acetone, methyl ethyl ketone, methyl isobutyl ketone, and denatonium.

In the United States, mixtures sold as denatured alcohol often have much greater percentages of methanol, and can be less than 50% ethanol.

Denaturing alcohol does not chemically alter the ethanol molecule. Rather, the ethanol is mixed with other chemicals to form a toxic or bad tasting solution. For many of these solutions, there is no practical way to separate the components.

Uses

Denatured alcohol is used identically to ethanol itself except for applications that involve fuel, surgical and laboratory stock. Regular ethanol is re□uired for food and beverage applications and certain chemical reactions where the denaturant would interfere. In molecular biology, denatured ethanol can be used for precipitating nucleic acids.

Purpose

In some countries, sales of alcoholic beverages are heavily taxed. In order to avoid paying beverage taxes on alcohol that is not meant to be consumed, the alcohol must be "denatured", or treated with added chemicals to make it unpalatable.

Denatured alcohol is not, in itself, a product that would be normally demanded if given the alternative of normal ethanol. Denatured alcohol and its manufacture are a public policy compromise. The supply and demand for denatured alcohol arises from the fact that normal alcohol (which in everyday language refers specifically to ethanol, suitable for human ingestion as a recreational drink or extractive medium for medicinal tinctures) is usually very expensive in comparison with similar chemicals, being highly taxed for revenue and public health policy purposes (see Pigovian tax). If pure ethanol were made cheaply available for fuel, solvents, or medicinal purposes, some people might ingest it.

Denatured alcohol provides a solution to permit industrial use and manufacture of ethanol, whereby cheap ethanol can be made available for non-consumption use without the risk of its being converted for consumption. The process creates an ethanol-containing solution that is not suitable for drinking, but is otherwise somewhat similar to ethanol for certain purposes. As a result, there is no duty on denatured alcohol in most countries, making it considerably cheaper than pure ethanol. As a conse□uence, its composition is tightly defined by government regulations that vary between countries.

Toxicity

Despite its poisonous content, denatured alcohol is sometimes consumed as a surrogate alcohol. This can potentially result in blindness or death if it contains methanol. For instance, during the American Prohibition, federal law re□uired methanol in domestically-manufactured industrial alcohols. On Christmas Day, 1926, and the two following days, which was roughly at the midpoint of the "Great Experiment" of nationwide alcohol prohibition, 31 people in New York City alone died of methanol poisoning. To help prevent this, denatonium is often added to give the substance an extremely bitter flavour. Substances such as pyridine are added to give the mixture an unpleasant odour, and agents such as syrup of ipecac may also be included to induce emesis.

BASICS OF PRODUCTION

Small scale production

Alcohol Fuel: A Guide to Making and Using Ethanol as a Renewable Fuel grassroots guide that gives readers all the information they need for making and using ethanol for fuel.

Ideally, your ethanol plant would be part of a farm or market-growing venture, for two reasons. First, as a grower you'd already have a familiarity with the day-to-day practices that agriculture entails. This includes working within a routine, searching for markets, dealing with e□uipment in both fair and inclement weather, and □uite importantly, improvising when necessary to keep things running smoothly. As anyone who has worked the land can tell you, the most successful farmers are well-rounded Renaissance people who can roll with the punches and take things in stride.

Second, a working farm provides a ready-made outlet for the manufactured fuel and its by-products. Most any internal-combustion engine or heating appliances can be adapted to run on alcohol this inventory includes tractors, trucks, pumps, generators, burners and furnaces — and the residual material from mash production contains enough nutrient to supplement normal livestock feed.

If agriculture is not in your background, it's still possible to manufacture alcohol, even economically, provided you have a reliable source of raw material, or feedstock. There are many viable candidates for ethanol production, including both sugar and starch crops. Residues from canning and juicing operations, even far from the farm, are also distinct possibilities. Realistically, it would be difficult to carry on much more than an experimental venture in a confined space such a suburban backyard, but it's still possible. Ideally, a rural setting or a location where there's room to expand and function without interference would be the better choice.

Flow Diagram for Ethyl Alcohol Production

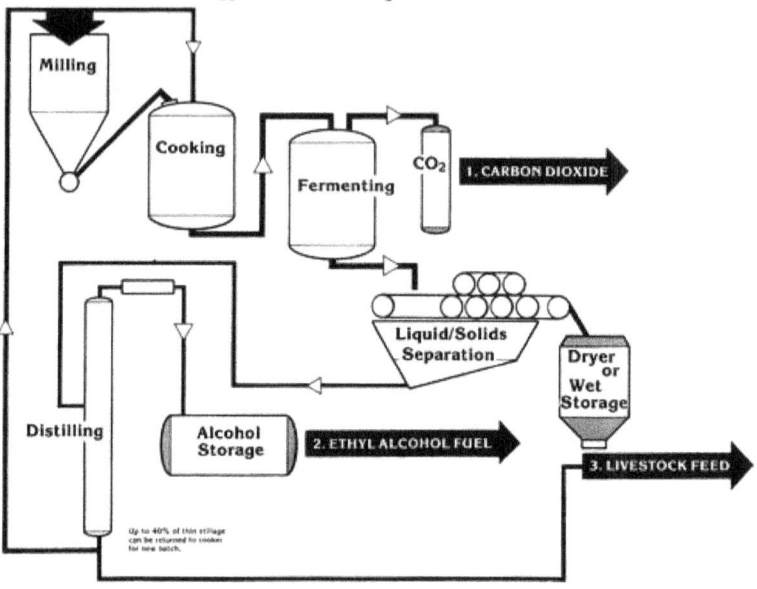

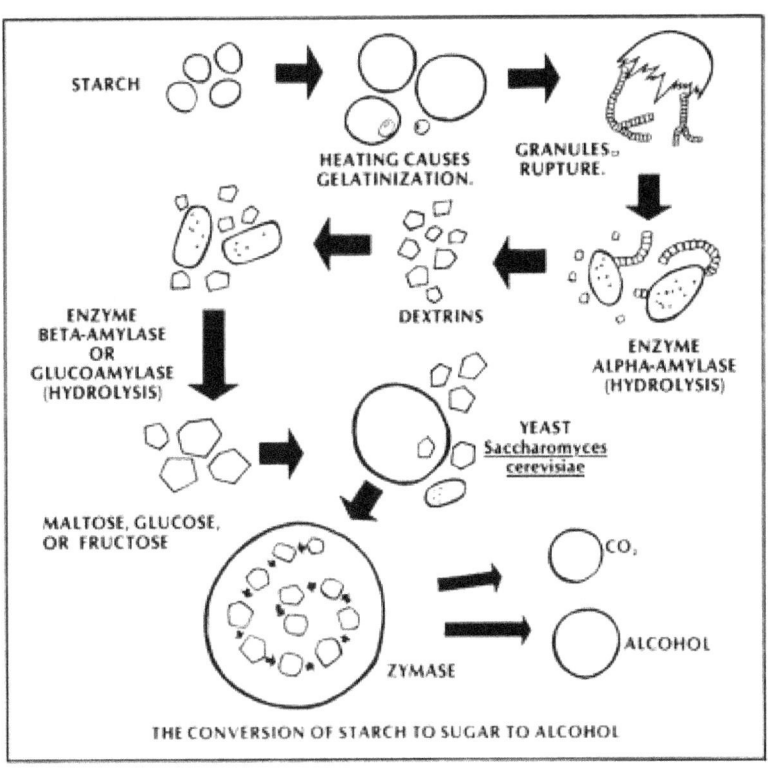

STARCH

HEATING CAUSES GELATINIZATION.

GRANULES RUPTURE.

DEXTRINS

ENZYME BETA-AMYLASE OR GLUCOAMYLASE (HYDROLYSIS)

ENZYME ALPHA-AMYLASE (HYDROLYSIS)

YEAST
Saccharomyces cerevisiae

MALTOSE, GLUCOSE, OR FRUCTOSE

CO_2

ALCOHOL

ZYMASE

THE CONVERSION OF STARCH TO SUGAR TO ALCOHOL

Buy It or Build It?

For the small-scale fuel producer, many still designs are so basic that it's much simpler and far less expensive to build the e☐uipment rather than to buy it. This is especially true of small-capacity operations. Costly stainless steel components aren't needed at this scale ordinary mild steel pipe will do for the columns and water lines, and in some applications plastic piping can be used. Likewise, tanks and vats need not be anything special, but for those elements, it's often cheaper to just buy used e☐uipment at a farm auction (stainless steel dairy storage and processor tanks are common auction items).

If you have welding skills and a place to work, you're way ahead of the game. For the kind of components involved, there's no real reason to use new materials. Any salvage or metal scrap yard is likely to produce the sort of parts you'll need. If you're not fussy, an old oil tank can make a decent boiler vat, and similar liquid storage containers can be adapted to serve as agitated mash cookers. Many components are make-do items from other applications, so you'll have to use a creative eye when shopping for good candidates. Unfortunately, many manufactured steel items particularly stainless steel have increased in value in the pre-owned marketplace because there is an increased foreign market for ☐uality steel salvage in general and for well-made American products in particular, especially among developing nations. The plumbing parts are for the most part standard off-the-shelf items.

Paying for the services of a professional welder will increase the cost of the e☐uipment considerably and perhaps even double it. You can trim expenses by locating all the materials yourself and preparing the parts to be fit and welded prior to delivering the job. The less the welder has to do in shaping, fitting and grinding, the less time he or she will spend on the project, reducing the hourly charge. This prep work is not a particularly high-skilled endeavor, and the investment in tools is very reasonable at this stage, so you might consider taking this approach and saving a few dollars in the bargain.

The Value of Your Time

Unfortunately for some of us, we are blessed with a desire to learn and accomplish rather than driven to make a profit. Such is the case for those working at the preliminary stages of setting up a home-scale distillery. Still, putting a lot of sweat equity into your ethanol project is a sound decision, especially for those who aren't fully committed to the idea of making large volumes of fuel alcohol. It reduces the amount of monetary investment involved (and thus the risk) and also provides you an intimate familiarity with the equipment that you'd never experience simply by purchasing it.

Once you're at the point of producing ethanol, you should place some value on your time, even if it is minimal. Assigning a cost per hour to your labor in collecting and processing feedstock, maintaining the distillery's operation, and handling the ethanol product and its record keeping will allow you to honestly and accurately calculate what it costs to be independent of the normal petroleum fuel network.

Calculating Cost Per Gallon

It is not that difficult to figure out what it will cost you to make a gallon of ethanol fuel, given some degree of stability in the cost of your cooking/heating fuel and feedstock sources. In a traditional farming operation, the costs of production are well established and independent of yield per acre and market value of the crop, until it comes time to calculate the actual level of profit.

The situation is similar with ethanol fuel, though many producers, particularly those working with spoilage and processing surplus, will not be concerned with crop yields other than their value in starch or sugars.

For example, if you've made 100 gallons of 185-proof ethanol in one run and 50 gallons of 190 proof ethanol in another, you can conclude that your yield is 140 gallons of 100 percent ethanol. The actual product, of course, is not that pure, but you're simply establishing a standard common denominator you can work with for the purposes of calculation.

Once that's established, you can determine the cost of your raw material feedstock, calculate the cost of transporting it to your work site, and subtract the value of any by-product yield, whether it's sold as distiller's grain or used for yourself at fair market value. This would include carbon dioxide for bottling, and any cellulosic co-products, which can be fermented to produce methane gas or dried for boiler fuel.

At this point you have a net feedstock value, for which you must now factor the cost of conversion to ethanol. The operating expenditures involved in this process include the cost of supplies such as enzymes and yeast, the cost of fuel to cook the mash and heat the distillation boiler, and the cost of insurance, licensing and any financing. These are added to the net feedstock figure to give you the cost of ethanol prior to adjustments for depreciation and other miscellaneous costs such as electricity for pumping, maintenance and repairs. Depreciation may be the cost of any leased equipment or machinery purchased, which can be extended or amortized over a given period, generally five years. Labor costs can also be considered here, though they may change with increased or decreased production.

The total fuel cost is then established by adding the adjusted costs above to the pre-adjusted cost of your ethanol to get a net cost. Dividing this figure by the number of pure ethanol gallons (not actual gallons) will give you the cost per gallon of your hard-earned product.

Beginning in 2005, an enhanced Small Producer Tax Credit became available with passage of the Energy Policy Act of 2005. Section 40 of the US Internal Revenue Code now allows an eligible small ethanol producer, defined as one manufacturing less than 60 million gallons per year, a federal income tax credit equal to $ .10 per gallon for the first 15 million gallons produced. Individual states may have other such incentives for small producers as well.

ETHANOL END USE

Ethanol is a high-quality, stable liquid. Some of the chemical and physical properties of ethanol are summarized in Table 1.

Property Ethanol

Chemical Formula.............................. $[C_2][H_5]OH$

Molecular Weight.............................. 46.07

Density (20[degrees] C)....................... 0.791 g/cc

Boiling Point [a]............................. 78.5[degrees] C

Heat of Combustion [b]........................ 5625 Kcal/1

Heat of Vaporization [c]...................... 9.225 Kcal/mole

Octane Rating................................. 106-108

Stoichiometric Air/Fuel Ratio [d]............. 9/1

[a] Boiling point is the temperature at which a liuid changes phase and becomes a gas; the point at which the vapor pressure of the liuid euals the vapor pressure of the system.

[b] Heat of combustion is the amount of heat given off when a unit uantity of any hydrocarbon (e.g., ethanol) is burned to carbon dioxide and water.

[c] Heat of vaporization is the heat input reuired to change liuid at its boiling point to a vapor at the same temperature (e.g., water at 100[degrees] C to steam at 100[degrees] C).

[d] The stoichiometric air/fuel ratio is the amount of air necessary completely to oxidize (burn) the fuel.

Common sources of sugar for production (sugar cane, jatropa, waste fruits, etc) and how to process them.

Ethanol may be produced from a variety of farm crops and wastes. The suitability of each type of feedstock may be assessed in terms of its calculated yield of ethanol, its availability by season and region of the U.S., and its cost.

TYPES OF FEEDSTOCK

Feedstock suitable for use in ethanol production via fermentation must contain sugars, starches, or cellulose that may readily be convertible to fermentable sugars. Feedstocks can be classified roughly into three groups: those containing predominantly sugars, starches, or cellulose, as shown below.

Sugars: sugar beets, sugar cane, sweet sorghum, ripe fruits

Starches: grains, potatoes, Jerusalem artichokes

Cellulose: stover, grasses, wood

The fermentation and distillation processes for two of the feedstock types, starch and sugar, are essentially identical. Their variations occur in storage requirements for the feedstock, the preparation of the fermentable sugar from the raw feedstock, and the" type of by-product produced.

The type of feedstock used has implications both for feedstock storage and in length of time during the year that an ethanol production plant could reasonably be expected to operate. Storage of any of the small grains would be the same whether they were to be used for feed or for alcohol production, i.e., moisture content, etc. would need to be controlled in order to prevent deterioration.

Sweet sorghum, sugar cane, and sugar beets have a short storage life in their harvest form.

Traditionally, the sugar industry has extended its processing season by extracting and storing the sugars in the form of molasses. The storage life of the feedstock is then considerably lengthened. Potatoes have approximately a six-month storage period prior to the start of any significant deterioration in their sugar/starch content.

Overripe or damaged fruits have an extremely short storage life and need to be processed ☐uickly. However, alcohol production from these materials aids in alleviating their disposal problem.

SUGAR CROPS : Preparation is basically a crushing and extraction of the sugars which the yeast can immediately use. But sugar crops must be dealt with fairly ☐uickly before their high sugar and water content causes spoilage. Because of the danger of such spoilage, the storage of sugar crops is not practical.

Sugar Cane. At the present time only 4 states (Florida, Louisiana, Texas, and Hawaii) cultivate sugar cane, but there are hybrids (such as saccharum spontaneum) which can be grown further north. High yields per acre of both sugar and crop residue are strong points of sugar cane production. The crop residue, called bagasse, is used in Brazil to provide heat for the distilleries.

Sugar Beets. Although sugar beets are grown in many areas of the U.S., they must be rotated with nonroot crops (1 beet crop per 4 year period is the general rule). While beet by-products cannot provide fuel for the distillery, the beet pulp and tops are excellent feed in wet or dry form. Or the tops may be left on the field for fertilizer and erosion control.

RAW MATERIALS

SUGAR CROPS

Interest in ethanol production from agricultural crops has prompted research on the development of sugar crops that have not been cultivated on a widespread commercial basis in this country. Three of the principal crops now under investigation are sweet sorghum, Jerusalem artichokes, and fodder beets.

Sweet Sorghum. Sweet sorghum is a name given to varieties of a species of sorghum: Sorghum bicolor. This crop has been cultivated on a small scale in the past for production of table syrup, but other varieties can be grown for production of sugar. The most common types of sorghum species are those used for production of grain.

There are two advantages of sweet sorghum over sugar cane: its great tolerance to a wide range of climatic and soil conditions, and its relatively high yield of ethanol per acre. In addition, the plant can be harvested in three ways: (1) the whole plant can be harvested and stored in its entirety;

(2) it can be cut into short lengths (about 4 inches long) when juice extraction is carried out immediately; and

(3) it can be harvested and chopped for ensilage. Since many varieties of sweet sorghum bear significant ⬜uantities of grain (milo), the harvesting procedure will have to take this fact into account.

The leaves and fibrous residue of sweet sorghum contain large ⬜uantities of protein, making the residue from the extraction of juice or from fermentation a valuable livestock feed. The fibrous residue can also be used as boiler feed

A member of the sunflower family, this crop is native to North America and well-adapted to northern climates. Like the sugar beet, the Jerusalem artichoke produces sugar in the top growth and stores it in the roots and tuber. It can grow in a variety of soils, and it is not demanding of soil fertility. The Jerusalem artichoke is a perennial; small tubers left in the field will produce the next season's crop, so no plowing or seeding is necessary.

Although the Jerusalem artichoke traditionally has been grown for the tuber, an alternative to harvesting the tuber does exist. It has been noted that the majority of the sugar produced in the leaves does not enter the tuber until the plant has nearly reached the end of its productive life. Thus, it may be possible to harvest the Jerusalem artichoke when the sugar content in the stalk reaches a maximum, thereby avoiding harvesting the tuber. In this case, the harvesting equipment and procedures are essentially the same as for harvesting sweet sorghum or corn for ensilage.

Fodder Beets. Another promising sugar crop which presently is being developed in New Zealand is the fodder beet. The fodder beet is a high yielding forage crop obtained by crossing two other beet species, sugar beets and mangolds. It is similar in most agronomic respects to sugar beets. The attraction of this crop lies in its higher yield of fermentable sugars per acre relative to sugar beets and its comparatively high resistance to loss of fermentable sugars during storage . Culture of fodder beets is also less demanding than sugar beets.

Fruit Crops. Fruit crops (e.g., grapes, apricots, peaches, and pears) are another type of feedstock in the sugar crop category. Typically, fruit crops such as grapes are used as the feedstock in wine production. These crops are not likely to be used as feedstocks for production of fuel-grade ethanol because of their high market value for direct human consumption. However, the coproducts of processing fruit crops are likely to be used as feedstocks because fermentation is an economical method for reducing the potential environmental impact of untreated wastes containing fermentable sugars.

STARCH CROPS

In starch crops, most of the six carbon sugar units are linked together in long, branched chains (called starch). Yeast cannot use these chains to produce ethanol. The starch chains must be broken down into individual six- carbon units or groups of two units. The starch conversion process, described in the previous chapter, is relatively simple because the bonds in the starch chain can be broken in an inexpensive manner by the use of heat and enzymes, or by a mild acid solution.

From the standpoint of ethanol production, the long, branched chain arrangement of six-carbon sugar units in starch crops has advantages and disadvantages. The principal disadvantage is the additional equipment, labor, and energy costs associated with breaking down the chain so that the individual sugar units can be used by the yeast. However, this cost is not very large in relation to all of the other costs involved in ethanol production. The principal advantage in starch crops is the relative ease with which these crops can be stored, with minimal loss of the fermentable portion. Ease of storage is related to the fact that a conversion step is needed prior to fermentation: many microorganisms, including yeast, can utilize individual or small groups of sugar units, but not long chains. Some microorganisms present in the environment produce the enzymes needed to break up the chains, but unless certain conditions (such as moisture, temperature, and pH) are just right, the rate of conversion is very slow. When crops and other feeds are dried to about 12% moisture -- the percentage at which most microorganisms cannot survive -- the deterioration of starch and other valuable components (for example, protein and fats) is minimal. There are basically two subcategories of starch crops: grains (e.g., corn, sorghum, wheat, and barley) and tubers (e.g., potatoes and sweet potatoes). The production of beverage-grade ethanol from both types of starch crops is a well established practice.

Much of the current agronomic research on optimizing the production of ethanol and livestock feed from agricultural crops is focused on unconventional sugar crops such as sweet sorghum. However, opportunities also exist for selecting new varieties of grains and tubers that produce more ethanol per acre. For example, when selecting a wheat variety, protein content is usually emphasized. However, for ethanol production, high starch content is desired. It is well known that wheat varieties with lower protein content and higher starch content usually produce more grain per acre and, conse□uently, produce more ethanol per acre.

CROP RESIDUE

The "backbone" of sugar and starch crops the stalks and leaves is composed mainly of cellulose. The individual six-carbon sugar units in cellulose are linked together in extremely long chains by a stronger chemical bond than exists in starch. As with starch, cellulose must be broken down into sugar units before it can be used by yeast to make ethanol. However, the breaking of the cellulose bonds is much more complex and costly than the breaking of the starch bonds. Breaking the cellulose into individual sugar units is complicated by the presence of lignin, a complex compound surrounding cellulose, which is even more resistant than cellulose to enzymatic or acidic pretreatment. Because of the high cost of converting liquefied cellulose into fermentable sugars, agricultural residues (as well as other crops having a high percentage of cellulose) are not yet a practical feedstock source for small ethanol plants. Current research may result in feasible cellulosic conversion processes in the future.

FORAGE CROPS

Forage crops (e.g., forage sorghum, Sudan grass) hold promise for ethanol production because, in their early stage of growth, there is very little lignin and the conversion of the cellulose to sugars is more efficient. In addition, the proportion of carbohydrates in the form of cellulose is less than in the mature plant. Since forage crops achieve maximum growth in a relatively short period, they can be harvested as many as four times in one growing season. For this reason, forage crops cut as green chop may have the highest yield of dry material of any storage crop. In addition to cellulose, forage crops contain significant quantities of starch and fermentable sugars which can also be converted to ethanol. The residues from fermentation containing nonfermentable sugars, protein, and other components may be used for livestock feed.

COPRODUCT YIELDS

Ethanol

The yield of ethanol from agricultural crops can be estimated if the amount of fermentable component sugar, starch, and cellulose is known prior to fermentation. If the yield is predicted based on percentages at the time of harvest, then the loss of fermentable solids during storage must be taken into account. This factor can be significant in the case of sugar crops, as discussed earlier.

The potential yield of ethanol is roughly one-half pound of ethanol for each pound of sugar. However, not all of the carbohydrate is made available to the yeasts as fermentable sugars, nor do the yeasts convert all of the fermentable sugars to ethanol. Thus, for estimating purposes, the yield of ethanol is roughly one gallon for each 15 pounds of sugar or starch in the crop at the time the material is actually fermented. Because of the many variables in the conversion of liquefied cellulose to fermentable sugar, it is difficult to estimate active ethanol yields from cellulose.

Carbon Dioxide

The fermentation of six-carbon sugars by yeast results in the formation of carbon dioxide as well as ethanol. For every pound of ethanol produced, 0.957 pound of carbon dioxide is formed; stated another way, for every 1 gallon of ethanol produced, 6.33 pounds of carbon dioxide are formed. This ratio is fixed; it is derived from the chemical equation:

Other Coproducts

The conversion and fermentation of agricultural crops yield products in addition to ethanol and carbon dioxide. For example, even if pure glucose is fermented, some yeast will be grown, and they would

represent a coproduct. These coproducts have considerable economic value, but, since they are excellent cultures for microbial contaminants, they may represent a pollutant if dumped onto the land.

Therefore, it becomes doubly important that these coproducts be put to good use.

Table 5.1 Average Alcohol Yield per Weight from Raw Material		
Feedstock Material	Yield in Gal./Cwt.	Yield in Gal./Ton
Wheat	4.25	85.0
Corn, Field	4.20	84.0
Buckwheat	4.17	83.4
Raisins	4.07	81.4
Sorghum grain	3.97	79.5
Rice	3.97	79.5
Barley	3.96	79.2
Dates, dry	3.95	79.0
Rye	3.93	78.6
Prunes, dry	3.60	72.0
Molasses, blackstrap	3.52	70.4
Sorghum cane	3.52	70.4
Oats	3.18	63.6
Figs, dry	2.95	59.0
Soybeans	2.44	48.8
Sweet Potatoes	1.71	34.2
Crabapples	1.29	25.8
Yams	1.36	27.3
Peanuts	1.35	27.0
Potatoes	1.14	22.9
Sugar Beets	1.10	22.1
Figs, fresh	1.05	21.0
J. Artichoke	1.00	20.0
Citrus Waste	0.83	16.6
Pineapples	0.78	15.6
Cranberries	0.78	15.6
Sugar Cane	0.76	15.2
Grapes	0.75	15.1
Apples	0.72	14.4
Apricots	0.68	13.6
Pears	0.57	11.3
Peaches	0.57	11.5
Plums	0.54	10.9
Pumpkins	0.49	9.8
Carrots	0.49	9.8
Whey	0.37	7.4

Note: Alcohol yields at 190 proof. Yields compiled from Cramer's EtOH Fuel Book (1980) and USDA sources. Short Hundredweight (Cwt.) equals 100 pounds US. Short Ton equals 2,000 pounds US.

Table 5.2 Average Alcohol Yield Per Acre from Raw Material	
Feedstock Material	Yield in Gal./Acre
Sugar Beets	287 - 412
Sugar Cane	268 - 335
Corn, Field	214 - 390
J. Artichoke	180 - 613
Potatoes	138 - 299
Sweet Potatoes	141 - 190
Apples	140
Dates, dry	126
Carrots	121
Raisins	102
Yams	94
Grapes	90
Peaches	84
Prunes, dry	83
Pineapples	78
Pumpkins	78
Cranberries	70
Rice	66 - 175
Soybeans	58
Pears	49
Barley	48 - 83
Molasses, blackstrap	45
Apricots	41
Peanuts	40
Oats	36 - 57
Sorghum grain	35 - 125
Buckwheat	34
Wheat	33 - 79
Figs, fresh	31
Figs, dry	29
Sorghum cane	26 - 500*
Rye	24 - 54
Plums	22
Citrus Waste	N/A
Whey	N/A
Crabapples	N/A

Note: Alcohol yields at 199.5 proof. Figures rounded to nearest whole number. Yields based on multiple sources including USDA Misc. Pub. 327, December 1938 and USDA Agricultural Statistics, 1938 and 2001.
* Lipinski, E.S., "Fuels From Sugar Crops," 1979

Source: Freudenberger, Richard. Alcohol Fuel: A Guide to Making and Using Ethanol as a Renewable Fuel (Books for Wiser Living from Mother Earth News). New Society Publishers (October 1, 2009).

Acid hydrolysis of starch is accomplished by directly contacting starch with dilute acid to break the polymer bonds. This process hydrolyzes the starch very rapidly at cooking temperatures and reduces the time needed for cooking. Since the resulting pH is lower than desired for fermentation, it may be increased after fermentation is complete by neutralizing some of the acid with either powdered limestone or ammonium hydroxide. It also may be desirable to add a small amount of glucoamylase enzyme after pH correction in order to convert the remaining dextrins.

High-temperature versus low-temperature cooking. Grain must be cooked to rupture the starch granules and to make the starch accessible to the hydrolysis agent. Cooking time and temperature are related in an inverse ratio: high temperatures shorten cooking time. Industry practice is to heat the meal-water mixture by injecting steam directly rather than by heat transfer through the wall of the vessel. The latter procedure runs the risk of causing the meal to stick to the wall; the subse|uent scorching or burning would necessitate a shutdown to clean the surface.

High-temperature cooking implies a high-pressure boiler. Because regulations may re|uire an operator in constant attendance for a high-pressure boiler operation, the actual production gain attributable to the high temperature must be weighed against the cost of the operator. If there are other supporting rationale for having the operator, the entire cost does not have to be offset by the production gain.

Continuous versus batch processes. Cooking can be accomplished with continuous or batch processes. Batch cooking can be done in the fermenter itself or in a separate vessel. When cooking is done in the fermenter, less pumping is needed and the fermenter is automatically sterilized before fermenting each batch. There is one less vessel, but the fermenters are slightly larger than those used when cooking is done in a separate vessel. It is necessary to have cooling coils and an agitator in each fermenter.

If cooking is done in a separate vessel, there are advantages to selecting a continuous cooker. The continuous cooker is smaller than the fermenter, and continuous cooking and hydrolysis lend themselves very well to automatic, unattended operation. Energy consumption is less because it is easier to use counterflow heat exchangers to heat the water for mixing the meal while cooling the cooked meal. The load on the boiler with a continuous cooker is constant. Constant boiler load can be achieved with a batch cooker by having a separate vessel for preheating the water, but this increases the cost when using enzymes.

Continuous cooking offers a high-speed, high-yield choice that does not re□uire constant attention. Cooking at atmospheric pressure with a temperature a little over 200 deg F (93 deg C) yields a good conversion ratio of starch to sugar, and no high-pressure piping or pumps are re□uired.

Separation versus nonseparation of nonfermentable solids. The hydrolyzed mash contains solids and dissolved proteins as well as sugar. There are some advantages to separating the solids before fermenting the mash, and such a step is necessary for continuous fermentation.

Batch fermentation re□uires separation of the solids if the yeast is to be recycled. If the solids are separated at this point, the beer column will re□uire cleaning much less fre□uently, thus increasing the feasibility of a packed beer column rather than plates. The sugars that cling to the solids are removed with the solids. If not recovered, the sugar contained on the solids would represent a loss of 20%o of the ethanol. Washing the solids with the mash water is a way of recovering most of the sugar.

FERMENTATION

Continuous fermentation. The advantage of continuous fermentation of clarified beer is the ability to use high concentrations of yeast (this is possible because the yeast does not leave the fermenter). The high concentration of yeast results in rapid fermentation and, correspondingly, a smaller fermenter can be used. However, infection with undesired micro-organisms can be troublesome because large volumes of mash can be ruined before the problem becomes apparent.

Batch fermentation. Fermentation time periods similar to those possible with continuous processes can be attained by using high concentrations of yeast in batch fermentation. The high yeast concentrations are economically feasible when the yeast is recycled. Batch fermentations of unclarified mash are routinely accomplished in less than 30 hours. High conversion efficiency is attained as sugar is converted to 10%-alcohol beer without yeast recycle. Further reductions in fermentation re uire very large uantities of yeast. The increases attained in ethanol production must be weighed against the additional costs of the e uipment and time to culture large yeast populations for inoculation.

Specifications of the fermentation tank. The configuration of the fermentation tank has very little influence on system performance. In general, the proportions of the tank should not be extreme. Commonly, tanks are upright cylinders with the height somewhat greater than the diameter. The bottom may be flat (but sloped for drainage) or conical. The construction materials may be carbon steel (commonplace), stainless steel, copper, wood, fiberglass, reinforced plastic, or concrete coated on the inside with sprayed-on vinyl. Usually, the tanks are covered to permit collection of the CO_2 evolved during fermentation so that the ethanol which evaporates with it can be recovered.

Many potential feedstocks are characterized by relatively large amounts of fibrous material. Fermentation of sugar-rich material such as sugar beets, sweet sorghum, Jerusalem artichokes, and sugarcane as chips is not a demonstrated technology and it has many inherent problems. Typically, the weight of the nonfermentable solids is equal to or somewhat greater than the weight of fermentable material. This is in contrast to grain mashes which contain roughly twice as much fermentable material as nonfermentable material in the mash. The volume occupied by the nonfermentable solids reduces the effective capacity of the fermenter. This means that larger fermenters must be constructed to equal the production rates from grain fermenters. Furthermore, the high volume of nonfermentable material limits sugar concentrations and, hence, the beer produced is generally lower in concentration (6% versus 10%) than that obtained from grain mashes. This fact increases the energy spent in distillation.

Since the nonfermentable solid chips are of larger size, it is unlikely that the beer containing the solids could be run through the beer column. It may be necessary to separate the solids from the beer after fermentation because of the potential for plugging the still. The separation can be easily accomplished, but a significant proportion of the ethanol (about 20%) would be carried away by the dewatering solids.

If recovery is attempted by "washing out," the ethanol will be much more dilute than the beer. Since much less water is added to these feedstocks than to grain (the feedstock contains large amounts of water), only part of the dilute ethanol solution from the washing out can be recycled through the fermenter. The rest would be mixed with the beer, reducing the concentration of ethanol in the beer which, in turn, increases the energy required for distillation.

Another approach is to evaporate the ethanol from the residue. By indirectly heating the residue, the resulting ethanol-water vapor mixture can be introduced into the beer column at the appropriate point. This results in a slight increase in energy consumption for distillation.

The fermenter for high-bulk feedstocks differs somewhat from that used for mash. The large volume of insoluble residue increases the demands on the removal pump and pipe plugging is more probable. Agitators must be sized to be self-cleaning and must prevent massive settling. High-speed and high-power agitators must be used to accomplish this.

The e□uipment for separating the fibrous residue from the beer when fermenting sugar crops could be used also to clarify the grain mash prior to fermentation. This would make possible yeast recycling in batch fermentation of grain.

Temperature control. Since there is some heat generated during fermentation, care must be taken to ensure that the temperature does not rise too high and kill the yeast. In fermenters the size of those for on-farm plants, the heat loss through the metal fermenter walls is sufficient to keep the temperature from rising too high when the outside air is cooler than the fermenter. Active cooling must be provided during the periods when the temperature differential cannot remove the heat that is generated. The maximum heat generation and heat loss must be estimated for the particular fermenter to assure that water cooling provisions are ade□uate.

Figure 1. Fermentation Vat Cooling Coil

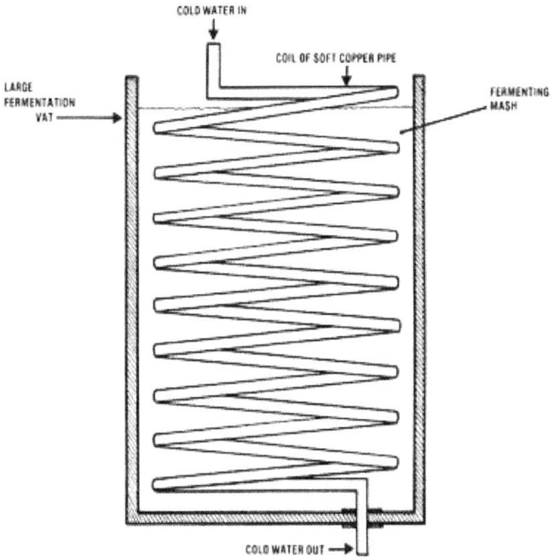

COLD WATER IN

COIL OF SOFT COPPER PIPE

LARGE
FERMENTATION
VAT

FERMENTING
MASH

COLD WATER OUT ➔

Fermentation Addendum

The optimum fermentation conditions are a temperature of 86 deg F (30 deg C) and a pH of 4 to 5. When the grains are left in the sweet wort or when backset is utilized, a buffering capacity is added that assists in maintaining the required pH (acidity). The expected alcohol yield from a 15-25% solution of fermentable sugars is 6.75 to 11.25% by weight.

The time reⅢuired to complete the fermentation is dependent upon the strain of yeast used. A variety of yeasts were tested for molasses fermentation in order to find a yeast strain that is highly efficient under variable conditions. (A group of 12 so tested is listed in the table below.) The ATCC 4132 produced 93 to 95% of the theoretical yield of alcohol from molasses without molasses pretreatment. The remainder of the yeasts were less efficient in alcohol production with the 48-hour fermentation efficiency ranging down to 35% (Heinz, September 11, 1979).

Yeast Strains and Their Relative Fermentation Efficiency

Yeast Strain	48-hr. Fermentation Efficiency (%)	Ethanol per Ton of Molasses (gallons)
ATCC 4132	93	73
CBS 1237	90	70
Y 7494	86	67
UCD 505	83	65
UCD 595	81	63
ATCC 26603	81	63
DADY	77	60
BAKER	77	60
ATCC 26602	62	48
NCYC 90	57	44
Y 2034	55	43
CBS 1235	35	27

Source: Freudenberger, Richard. Alcohol Fuel: A Guide to Making and Using Ethanol as a Renewable Fuel (Books for Wiser Living from Mother Earth News). New Society Publishers (October 1, 2009).

Source of yeast:

ATCC -- American Type Culture Collection
CBS -- Centraalbureau voor Schimmelcultures, The Netherlands
Y -- Northern Region Research Center, USA
UCD -- University of California, Davis
DADY -- Universal Foods Corporation
BAKER -- Local procurement
NCYC -- National Collection of Yeast Culture, Brewing Research
Foundation, England.

For small-scale production, the most readily available yeast is active dry yeast especially designed for distillers' use in grain mash fermentation. This product has been found to work well for beet, cane, and citrus molasses fermentation. The yeast is designed to produce uniform, rapid fermentation and maximum alcohol yields under a wide range of temperatures and pH. The time required for fermentation will vary with the temperature, although most estimates are for 48 to 72 hours.

Yields of alcohol may be reduced if there is any contamination of the sweet wort. Contamination with undesirable micro-organisms will decrease the yield of alcohol as these will compete with the yeast for the sugar. Prior to the addition of the yeast, contamination from external sources or from the equipment itself may occur readily in the cooling of the sugar mixture.

At this point in the process, the sugar solution is a suitable medium for growing a variety of microbes that may be pathogenic or produce toxic substances (Crombie, 1979). Microbes may be introduced with the raw materials initially, via the addition of cooling water, or from the air; thus, provision needs to be made for high-quality water, and the design must incorporate some protection from possible contamination from the air.

Fortunately the contamination problem is mitigated by the fact that yeast populations grow quite rapidly, and overwhelm many of the potentially competing organisms. In addition, the initial inoculation introduces a large yeast population that allows the yeast a head start. Provided care is exercised and, thus, unwanted microbial action does not occur, decreases in the yield of ethanol resulting from competing reactions can be held to a minimum.

Undesirable microbial reactions occurring in the fermentation step may produce unwanted substances in the stillage, but since these cannot be predicted in advance and would be batch-specific, thcy would need to be handled on an individual batch basis.

The solution to the contamination problem involves the design of the production facility and the training of the process operator. The plant must be designed in such a fashion that sanitation is readily accomplished and contamination of the sweet wort may be avoided when reasonable care is exercised. Once the operating procedures for a plant are established, these problems should be minimal.

All e□uipment currently being marketed utilizes a batch fermentation process; however, continuous fermentation units have been used in some industrial applications. A continuous fermentation process allows the use of smaller fermenters and supplementary equipment; it has been of interest for many years.

Continuous fermentation methods have been used successfully on waste sulfite liquor in Europe. Since sulfite li□uors are sterile and even antiseptic, continuous fermentation is possible and desirable.

With fermentable substances such as molasses, however, any contamination is cumulative and soon spreads throughout the system to reduce yields. The contamination can be controlled to some extent using penicillin or other antibiotics, but their use to control organisms competing for sugars and decreasing alcohol yields creates another problem: the use of by-product feed. The antibiotic content of feeds for livestock is carefully controlled by the Food and Drug

Administration. FDA requires tests to show that antibiotics and their degradation products in the resulting by-product feeds are below the maximum allowable levels. Most of these antibiotics are destroyed in the drying process. One commercial ethanol producer looked at and experimented with a continuous process, but finally gave up because of the contamination problems involved.

However, continuous fermentation could conceivably work well if the fermentable solution could be sterilized. But with grains or cellulose feedstocks, the grain particles and fiber present in the sweet wort make such sterilization extremely difficult.

Although continuous fermentation offers a more rapid method of producing ethanol with smaller tanks, etc., problems remain to be worked out before the system is feasible for the small-scale operator using grains as a feedstock. Work is underway at the present time to develop continuous automated e□uipment for small-scale ethanol production.

If wheat is used as the feedstock, special provisions must be made for the additional foaming that occurs during fermentation because of the presence of the gluten protein. Three possibilities exist to handle this problem: increasing the capacity of the equipment over that for the same quantities of corn; using a defoaming agent; or removing the gluten protein prior to the fermentation process.

Since the fermentation process produces heat and the optimum fermentation temperature is about 90 deg F (32 deg C), cooling is necessary in order for the yeast to survive and work efficiently. The formation of ethanol is accompanied by approximately 287 kilocalories per kilogram of ethanol produced: 517 BTU per pound or 3,418 BTU per gallon (Alfa-Laval, undated). If insufficient cooling is provided, the fermentation times are increased. Where no provision is made for removal of the heat of fermentation, heat losses may occur in both of these ways: from the evolution of carbon dioxide, and from convection and radiation from the walls and other surfaces of the fermenter vessel. Heat removal in the off-gas is relatively small, even though the gas is saturated with water vapor and its attendant evaporative cooling effect. If heat evolution is too great to be dissipated by radiation, the increasing temperature of the contents results in a decreased yeast activity and a greater heat release.

Table 1. Average yield of 99.5 percent alcohol per ton**

Material	Gallons	Material	Gallons
Wheat (all varieties)	85.0	Yams	27.3
Corn	84.0	Potatoes	22.9
Buckwheat	83.4	Sugar beets	22.1
Raisins	81.4	Figs, fresh	21.0
Grain sorghum	79.5	Jerusalem artichokes	20.0
Rice, rough	79.5	Pineapples	15.6
Barley	79.2	Sugarcane	15.2
Dates, dry	79.0	Grapes (all varieties)	15.1
Rye	78.8	Apples	14.4
Prunes, dry	72.0	Apricots	13.6
Molasses, blackstrap	70.4	Pears	11.5
Sorghum cane	70.4	Peaches	11.5
Oats	63.6	Plums (nonprunes)	10.9
Figs, dry	59.0	Carrots	9.8
Sweet potatoes	34.2		

Source: Freudenberger, Richard. Alcohol Fuel: A Guide to Making and Using Ethanol as a Renewable Fuel (Books for Wiser Living from Mother Earth News). New Society Publishers (October 1, 2009).

Table 2. Average yield of 99.5 percent alcohol per acre **

Material	Gallons	Material	Gallons
Jerusalem artichokes****	1200.0	Grapes (all varieties)	90.4
Sugarcane		Peaches	84.0
(Hawaii, 18 to 22 months)	889.0	Barley***	83.0
Sugar cane (Louisiana)***	555.0	Prunes, dry	82.8
Sorghum cane***	500.0	Wheat (all varieties)***	79.0
Sugar beet***	412.0	Pineapples	78.0
Potatoes***	299.0	Oats***	57.0
Corn***	214.0	Rye***	54.0
Sweet potatoes***	190.0	Pears	49.3
Rice, rough***	175.0	Molasses, blackstrap	45.0
Apples	140.0	Apricots	41.0
Dates, dry	126.0	Buckwheat	34.2
Grain sorghum***	125.0	Figs, fresh	31.5
Carrots	121.0	Figs, dry	29.5
Raisins	101.7	Plums (nonprunes)	21.8
Yams	94.0		

Source: Freudenberger, Richard. Alcohol Fuel: A Guide to Making and Using Ethanol as a Renewable Fuel (Books for Wiser Living from Mother Earth News). New Society Publishers (October 1, 2009).

DISTILLATION PROCESS

Distillation is a widely used method for separating mixtures based on differences in the conditions required to change the phase of components of the mixture. To separate a mixture of liquids, the liquid can be heated to force components, which have different boiling points, into the gas phase. The gas is then condensed back into liquid form and collected. Repeating the process on the collected liquid to improve the purity of the product is called double distillation. Although the term is most commonly applied to liquids, the reverse process can be used to separate gases by liquefying components using changes in temperature and/or pressure.

How the Distillation Process Works

Though there are many different designs used for alcohol-producing stills, every installation operates on the same set of principles. These general theories of distillation are impressively complicated, but fortunately once you understand a few of the basics, you should know enough to design and build your own ethanol plant.

Distillation is the separation of a liquid from other liquids or solids. Because each substance has a fixed rate of vaporization (which varies with heat) determined by the pressure the vapors develop in a closed container to achieve equilibrium with the fluid one liquid can be separated from other matter by carefully controlling the heat applied to the mixture. Alcohol's vapor pressure happens to be higher than water's, so ethanol's vapor pressure reaches an equilibrium with atmospheric pressure (the point at which a liquid boils) before water's vapor pressure does.

But when water and alcohol are mixed, the boiling point of the combination falls between the boiling points of the separate constituents (water will boil at 100 deg C; alcohol boils at 78.3 deg C). It is the ratio of the water to alcohol which determines the actual temperature of boiling for the mixture. More alcohol lowers the boiling point and less raises it -- so you can see that the temperature of the mash will rise throughout the distillation run as the alcohol is drawn off.

Because alcohol has a higher vapor pressure than water, the vapors given off by boiling a combination of the two will have a disproportionately large share of alcohol. For example, in a mash that has 10% alcohol and 90% water, the vapors released will be about 80% alcohol. To increase that percentage (and raise the proof), the vapors must be condensed and revaporized. Each redistillation raises the proof of the batch further until the li□uid reaches an azeotropic condition at 95.57%. The process of enrichment (or rectification) is halted by the balance (azeotropy) of alcohol and water in the vapors.

During simple pot distillation, the proof of the product at the beginning of the run is high ... but as the proof of the mash drops, the proof of the distillate also declines. In fact, the depletion in proof strength is geometric.

Moonshiners can manage to increase proof strength by adding "thumper kegs" or doublers to their stills. This involves running a line from the pot down into a secondary barrel before it continues to the condenser.

Vapors from the pot condense in the doubler and raise the heat of low-proof distillate in the bottom of the tank to the boiling point. High-proof alcohol vapors are then released. Several doublers can be added in series to boost alcohol content.

Though "thumper kegs" are sound in principle and do raise proof, they are very energy-inefficient. Still designers discovered that the enrichment process could be more effectively accomplished by stacking one still atop another. This technique is called the pipe column.

The first column still was a form of what is called "batch run bubble cap plate" design. The pipe column was divided into sections by plates, each of which had a hole in the middle with a short section of pipe (known as a riser) extending upward into the column directly over the hole. An inverted cup or cap was placed above the riser so that it didn't block the pipe's opening. Then another pipe (called a downcomer) was added, extending from a half-inch above one plate, through the next plate up, and ending one inch above that plate. Eight or more of these plates were used in a still.

Before operation, the column was filled with beer so that each plate was covered to the top of the downcomer. When heated, the vapors would rise from the bottom plate and be forced into the liquid above the next plate by the caps. The heat transferred to the li□uid by condensation raised the temperature of that level's fluid to boiling, so that a higher-grade vapor was emitted. By the time the vapors reached the top of an eight-or-more-tiered column, the proof was very high. While the vapors rose, the distilled water descended through the downcomers. Hence the name "countercurrent stream" was developed. The countercurrent method was basically just a more efficient simple pot with doubler design which still suffered from rapidly declining proof toward the end of the run.

The next design innovation was the development of a continuous-run still, which could take off a steady proof product throughout the run. Early designs had the mash introduced after being preheated to near saturation point midway in the column. At the bottom of the column a reboiler was used to add pressure and heat to the system. These two sources of heat served to e□ualize the distillation conditions throughout the column. Each plate had the right amount of heat for the percentage of water and alcohol present.

As beer was added to the column, the alcohol vaporized (along with a little water) and rose to the next plate. At this point the water (with a little alcohol) was stripped off and descended to the lower plate. By the time the water fell all the way to the bottom plate, any alcohol that could be released by distilling was on its way to the top of the pipe. With this method, distillation could be maintained indefinitely by adding additional feed at the entry in mid-column.

Today this system has been developed into three basic designs of e□uilibrium stills: the packed column, the perforated plate, and the bubble cap plate. All three work by the pre-established principles of enrichment and countercurrent flow.

Fermentation and distillation equipment

Packed column

The easiest equilibrium still to design and build is the packed column. Its components are a firebox, a "pot" (which is a tank of some sort), a pipe packed with a material which will leave 60-90% air space, and a condenser.

Here's an example of how such a still might be constructed: Find a 100-gallon tank and build a firebox under it. On top of the pot, make a port -- to allow access for cleaning and loading -- and then weld on a two-foot length of 5" pipe. Atop this tubing, add a reducer and three feet of 2" pipe. Pack this section with a material such as brass, copper, or stainless steel wool. Then reduce the column to 3/4" for the condenser. A still of this design provides 180-proof alcohol for about two-thirds of the run, and then proof dwindles.

There are a number of advantages to the packed column design. Improvements can yield proof of about 190, and the still can be run either continuously or on a batch basis. On a small scale, packed columns are inexpensive to build and quite easy to operate. However, on a large scale the design presents problems. In order to run continuously, the mash would have to be free of solids to avoid accumulations in the pot. But batch running would be very fuel-inefficient in a large-scale packed column still.

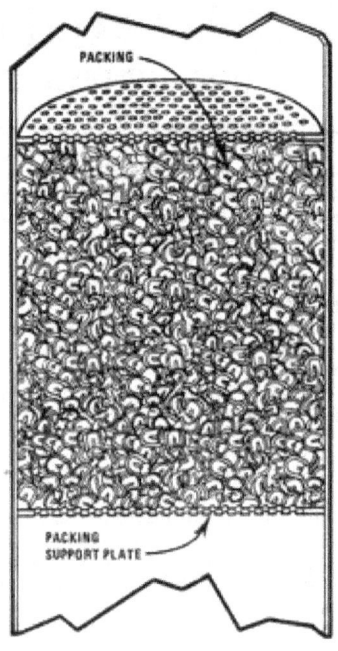

Perforated Plate

For continuous-run applications, it's hard to beat a perforated plate column. Because the construction is simpler than a bubble cap plate and involves only drilling or punching holes in the plates, it tends to be less expensive than the other plate design. Vapors flow through holes in the plates and are cooled by li☐uid flowing across the plates. Alcohol stays vaporous and rises, while the water sinks through downcomers.

There is one significant disadvantage to perforated plate stills, though. A minimum pressure must be maintained in the still or the liquid on the plate will "dump". Dumping occurs when the pressure holding the li☐uid on the plate drops far enough to allow the liquid to fall through the holes and down to the next lower plate. This stops distillation, and allows undistilled mash to escape through the spent mash drain. Thus, the perforated plate design presents certain problems for wood or biomass fuel systems where heating tends to vary.

Start with two 16' pieces of 12" thin-wall tubing and make 18 plates (from 1/8" metal) the same diameter as the inside of this pipe. Drill 1/2" holes in each plate to occupy about 8-10% of the surface area -- roughly

50-57 holes. Place these plates about 10-1/2 inches apart in one tube to form the stripper column. Then take 24 more plates and drill 490-520 holes (5/32" in diameter) in each one. These plates are spaced 7-1/2 inches apart in their 16' tube to form the rectification column. Each plate also has a 1-1/8"-diameter downcomer with a seal cup on the bottom. Each downcomer extends 1-1/8 inches above its plate.

A 4' thin-wall tube leads from the top of the stripper to the bottom of the rectification chamber. (It takes additional pressure to make the vapor flow downward, so ideally the stripper and rectification columns would be combined in one length. However, this makes the still impractically tall.) Another 4" pipe leads from the top of the rectification chamber to the condenser.

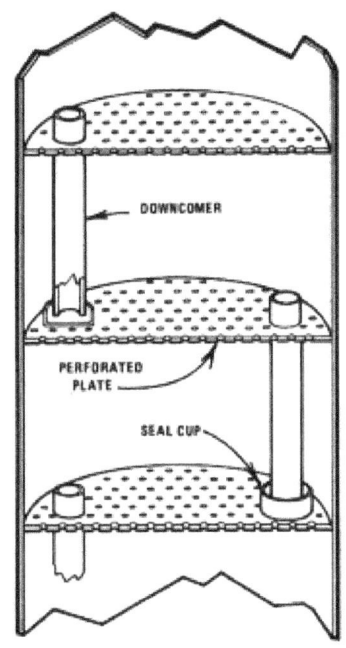

Bubble cap plate

Mash is pumped through the condenser for preheating, and is dropped into the top of the stripper column. The spent mash and solids are pumped out the bottom. Alcohol -- at about 100 proof -- leaves the top of the stripper column and enters the bottom of the rectification tower. Alcohol leaves the top at about 190 proof, and water (with about 0.05% alcohol) drains from the bottom.

Heat is provided by a steam generator or boiler and is introduced live or through coils at the bottom of the stripper column in the section called the "reboiler".

Bubble Cap Plate

The bubble cap plate distillation column is the oldest design still in use -- it is a derivation of the principle described at the beginning of this section. There are some obvious limitations for do-it-yourselfers in this system. Mash would have to be very free of solids to avoid clogging downcomers and caps, the risers, and the plates themselves. Such a still would probably re□uire the construction of ports for cleaning between the stages of the column.

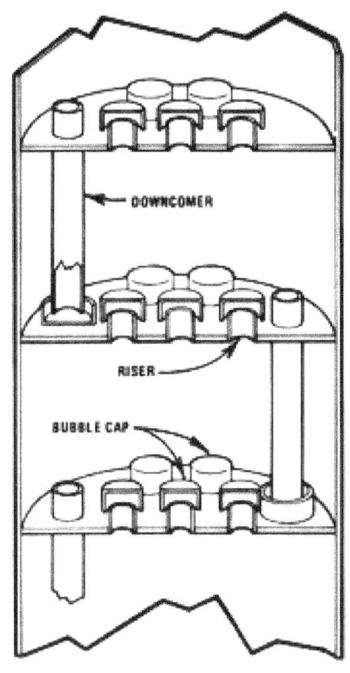

Solar Stills

In certain areas of the country, solar energy is a viable source of heat for distillation. In other areas it may not be worth the time to even think about it. The advantages seem obvious: solar energy is free, unending, and nonpolluting. However, the disadvantages seem to outweigh the advantages unless one is in an area where the sun can be counted on to shine. For example, the mash has to be run when it has finished the fermentation process, or it will turn to acetic acid within a few days. There are certain chemicals that can be added to the mash to hold it for a while, but doing so only adds to the cost of fuel production.

A solar still works best if the mash is preheated first. A serpentine pattern copper pipe solar panel to preheat the mash before it reaches the still can raise the mash temperature to anywhere from 120 to 180 deg F (48.8-82.2 deg C), depending on the s☐uare footage of the panel. Then, as the hot mash enters the still, it will easily flash to a vapor. Remember that water is taken with the alcohol, so this is not more than about 90- to 100-proof vapor. Also, considerable alcohol is left in the discharge, and that will have to be recirculated to remove all the alcohol from the mash. The second running will have an even lower proof.

Alcohol vapors rise, so -- to prevent adding more water to the distillate -- it's best to provide a means of removing the alcohol at the top of the still, not the bottom. If the condensed alcohol vapors are allowed to run down the glass, they will gather the water droplets that have condensed on the lower portion of the glass and thus reduce the proof.

ALCOHOL AS FUEL ENGINE

Before you begin to convert your automobile or truck engine to use alcohol, it's important that you understand the properties of and the differences between the two fuels.

Gasoline is a complex mixture of hydrocarbons ... substances comprising just hydrogen and carbon atoms. These hydrocarbons can appear in all forms (as a gas, liquid, or solid), but for our purposes, we're concerned with the fuel in its liquid state.

To derive various hydrocarbon fuels, the industry merely refines crude oil (made many millions of years ago as a result of geological and biological cycles) and draws off the desired product at a certain temperature and pressure. Hence there are the lighter, gaseous fuels such as butane, propane, and ethane ... the liᗺuids like octane, pentane, and hexane ... the heavier, oily liᗺuids such as kerosene and fuel oil ... and so on all the way down through waxes and finally solids.

Gasoline as we know it is a combination of octane, benzene, toluene, various other aromatics, tetraethyl lead, detergents ... and compounds of sulfur, phosphorus, and boron. Because of this complex mixture of ingredients - and because the refineries vary the blend to suit climate, seasonal changes, and altitude - it's difficult to choose a "representative" sample of gasoline for comparison purposes. Nonetheless, the figures that are given in the "Properties of Gasoline, Ethanol, and Methanol" chart which follows are fairly typical of average high-test automotive gasoline.

Alcohol, on the other hand, has to be manufactured ... in our case through fermentation and distillation processes. Because of the steps involved in its manufacture, alcohol has always been more expensive than gasoline to produce. But now, with dwindling crude oil supplies, the price of gasoline is skyrocketing ... and soon gasoline itself will probably have to be synthetically manufactured, at a cost far greater - since the production process is much more complicated than that of alcohol.

Alcohol compounds are also hydrocarbons ... but in alcohol, one of the hydrogen atoms has been supplanted by a hydroxyl radical (hence the OH symbol), which is an oxygen atom bonded to a hydrogen atom. Alcohols, too, take many forms and have various levels of complexity, but we're concerned mainly with ethanol (grain-derived alcohol) and - just in passing - methanol (wood- or cellulose-derived alcohol).

These two alcohols are the only practical alcohol fuels ... and of the two, ethanol is more economically feasible on a small scale. (The raw material used to make methanol - wood chips, garbage, or cellulose matter - is relatively inexpensive, but the manufacturing process necessary to produce methyl alcohol is economical only on an industrial level.)

On the surface, the difference between alcohol and gasoline might appear relatively minor: Alcohol contains oxygen, while gasoline doesn't. In reality, however, the dissimilarities are far more complex than that. Additionally, under compression - as is the case in an engine's combustion chamber - things get even more complicated ... but we'll get more into detail on these points later.

Regardless of the inherent differences between gasoline and alcohol, though, the fact is that alcohols make ideal motor fuels. The first practical internal combustion engine - patented by Nikolaus Otto in 1877 - ran on alcohol (gasoline had not been "discovered" yet), and the Model A Ford, produced from 1928 to 1931, was designed to burn a variety of fuels ... alcohol being one of them. In addition, Studebaker trucks built for export in the 1930's (and various domestic tractors sold both in the U.S. and abroad) were offered with either gasoline or alcohol fuel systems. (Indeed, at the start of the "motorized era", alcohol was just as common as - if not more so than - fossil fuels. But as time went on, the petroleum industry - which was organized and thus more powerful than the independent, often farm-based alcohol producers - lobbied successfully for the wholesale use of "superior" gasoline fuels. Strangely enough, in areas where petroleum had to be exclusively imported, or during time of war when gasoline supplies were rationed, alcohol suddenly became an excellent motor fuel again ... and was touted as such by the petroleum distributors who were selling it!)

Be that as it may, alcohol has characteristics that make it a natural engine fuel: It has a high "octane" rating, which prevents engine detonation (knock) under load, it burns clean ... so clean, in fact, that not only are noxious emissions drastically reduced, but the internal parts of the engine are purged of carbon and gum deposits ... which, of course, do not build up as long as alcohol is used as fuel, an alcohol burning engine tends to run cooler than its gasoline-powered counterpart, thus extending engine life and reducing the chance of overheating.

At this point, we can detail exactly how these and other characteristics of alcohol affect engine performance.

"OCTANE" RATING

Actually, when referring to alcohol fuels, the word "octane" does not apply, since octane (in its pure form) is merely the hydrocarbon in gasoline which is assigned the numerical value of 100 for fuel-rating purposes. The octane number given automotive fuels is really an indication of the ability of the fuel to resist premature detonation within the combustion chamber. (Premature detonation, or engine knock, comes about when the fuel/air mixture ignites spontaneously toward the end of the compression stroke because of intense heat and pressure within the combustion chamber. Since the spark plug is supposed to ignite the mixture at a slightly later point in the engine cycle, pre-ignition is undesirable, and can actually damage or even ruin an engine.)

Because a high compression ratio in an engine results in more power per stroke, greater efficiency, and better economy, it's easy to see why a fuel that resists pre-ignition even under high compression conditions is especially desirable ... and alcohol is, on the average, about 16 points higher on the research octane scale than premium gasoline.

HEAT VALUE

The heating value of a fuel is a measure of how much energy we can get from it on a per-unit basis, be it pounds or gallons. When comparing alcohol to gasoline using this "measuring stick", it's obvious that ethanol contains only about 63% of the energy that gasoline does ... mainly because of the presence of oxygen in the alcohol's structure. But since alcohol undergoes different changes as it's vaporized and compressed in an engine, the outright heating value of the ethanol isn't as important when it's used as a motor fuel.

The fact that there's oxygen in the alcohol's structure also means that this fuel will naturally be "leaner" in comparison to gasoline fuel without making any changes to the jets in the carburetor. This is one reason why we must enrich the air/fuel mixture (add more fuel) when burning alcohol by increasing the size of the jets, which we'll discuss further in another section.

VOLATILITY

The volatility of a fuel refers to its ability to be vaporized. This is an important factor, because if vaporization doesn't occur readily, the fuel can't be evenly mixed with air and is of little value in an engine. Some substances that are highly volatile can't easily be used as a motor fuel ... and others, which have excellent heating value, aren't volatile enough to be used in an engine (such as tars and waxes).

Another point to keep in mind is that a very volatile fuel is potentially dangerous, because of the chance of explosion from heat or sparks. This is one reason why alcohol, with a higher flash point than gasoline, is a much safer automotive fuel ... especially considering that the average car's storage tank is really ꝏuite vulnerable.

LATENT HEAT OF VAPORIZATION

Latent heat of vaporization is the phenomenon that results in an alcohol-powered engine's running cooler than its gasoline-fueled counterpart. When a substance is about to undergo a change in form (from a liquid to a vapor, in this case), it must absorb a certain amount of additional heat from its surroundings in order for the change to take place. Since alcohol must absorb roughly 2-1/2 times the amount of heat that gasoline does, and the heat naturally is taken from the engine block, the engine should operate at a much lower temperature ... in theory, that is.

What happens in reality is that the alcohol/air mixture doesn't have time to absorb all the heat it could during its short trip through the engine manifold. So instead of running 2-1/2 times cooler on alcohol than it does on gasoline (which, by the way, would not be desirable ... since an engine must retain a certain amount of heat to run efficiently), the engine operates at temperatures only slightly cooler - about 20-40 deg F lower, depending on the specific engine when using alcohol fuel.

EXHAUST EMISSIONS

When gasoline is burned in an engine, it produces carbon monoxide and other poisonous fumes ... mostly because of the fact that the fuel never combusts completely, and also because it's subjected to extreme temperatures and pressures. In addition, as we mentioned before, gasoline is a complex mixture of many substances ... and some of those substances are lead, sulfur, and other noxious materials. These, too, add to the contaminative effects of the engine's exhaust fumes.

Alcohol, on the other hand, burns much cleaner. Even though it, too, never combusts completely, the volume of noxious fumes is drastically reduced in an alcohol-burning engine ... because alcohol contains oxygen in its structure (which means more thorough combustion) but doesn't contain all the other pollutants necessary as additives in gasoline.

With alcohol fuel, however, the test results improved enormously. Even with all pollution controls removed from the engine (except for the PCV valve), the cab registered a mere 0.08% CO and only 25 PPM of HC ... the equivalent of 95% less CO and 87.5% less HC, or a total of about 92% cleaner!

ALCOHOL/WATER MIX

As we all know (some of us from experience), water and gasoline don't mix. The gasoline tends to float to the top of the mixture, leaving the water to settle below it. In a car's fuel tank, this can be disastrous, particularly during the winter season.

Alcohol, however, mixes quite well with water: The water particles distribute evenly within the mixture. As a result, not only is the winter freezing problem solved, but pure alcohol is not necessary for fuel purposes. This is very important to the small-scale alcohol fuel producer, since nonindustrial stills are generally not capable of producing more than 192-proof (96% pure) alcohol.

ENGINE ECONOMY

The fuel economy of an engine is directly proportional to how rich the air/fuel mixture is ... and that, of course, is dependent upon how large the main jet in the carburetor is. Alcohol requires a richer air/fuel mix than does gasoline (9-to-1 as opposed to 15-to-1), but that difference is not reflected proportionately with respect to economy ... partially due to the fact that alcohol has a higher "octane" rating and can be utilized more efficiently.

ENGINE PERFORMANCE

An engine powered by alcohol - if converted correctly - will have performance equivalent to, if not greater than, the same powerplant burning gasoline. This is because of the fact that alcohol has a higher "octane" rating (hence the timing can be advanced slightly), and it can stand much greater compression ratios.

Table 9.1 Liquid Fuel Characteristics of Gasoline and Ethanol		
Chemical Properties Formula	Gasoline complex mixture	Ethanol
Molecular Weight	variable	46.07
Percent Carbon (by weight)	85-88	52.14
Percent Hydrogen (by weight)	12-15	13.12
Percent Oxygen (by weight)	variable	34.74
Carbon/Hydrogen Ratio	5.6 - 7.4 : 1	4.0 : 1
Stochiometric Ratio (air-fuel)	14.2 - 15.1 : 1	9.0 : 1
Physical Properties	Gasoline	Ethanol
Specific Gravity	.70 - .78	.7936
Liquid Density (Lb./Cu. ft.)	43.6	49.3
Liquid Density (Lb./Gallon)	5.8 - 6.5	6.59
Boiling Point (Degrees F.)	80 - 440	173.3
Freezing Point (Degrees F.)	minus 70	minus 174.6
Solubility (in water)	240 ppm	miscible
Solubility (water in)	88 ppm	miscible
Vapor Pressure at 1 Bar (100°F.)	7 - 15 In. Hg.	2.5 In. Hg.
Vapor Pressure at 1 Bar (77°F.)	.3 In. Hg.	.85 In. Hg.

Table 9.1 cont. Liquid Fuel Characteristics of Gasoline and Ethanol		
Thermal Properties Formula	Gasoline complex mixture	Ethanol
Heat of Combustion (77°F.)		
Lower Heating Value (Btu/lb.)	18,900	11,550
Lower Heating Value (Btu/gal.)	115,400	76,114
Higher Heating Value (Btu/lb.)	20,250	12,780
Higher Heating Value (Btu/gal.)	124,800	84,220
Latent Heat of Vaporization (77°F. at 1 Bar) (Btu/lb.) (Btu/gal.)	150 900	395 2,603
Flash Point (Degrees F.)	minus 50	55
Autoignition Temperature (Degrees F.)	430 - 500	793
Octane Rating (Research)	90 - 101	106
Optimum Air-Fuel ratio	15 : 1	9 : 1
Explosive Limits Air-Fuel Ratio	13.2 : 1 - 71.4 : 1	5.8 : 1 - 23.3 : 1
Explosive Limits in Air (by percentage)	1.4 - 7.6	3.3 - 19
Maximum Practical Compression Ratio (spark ignition)	9.2 : 1	15 : 1

Source: Freudenberger, Richard. Alcohol Fuel: A Guide to Making and Using Ethanol as a Renewable Fuel (Books for Wiser Living from Mother Earth News). New Society Publishers (October 1, 2009).

www.ingramcontent.com/pod-product-compliance
Lightning Source LLC
Chambersburg PA
CBHW072143170526
45158CB00004BA/1484